Goat Production Manual
Second Edition

A Practical Guide

Marion (Meg) Smart DVM, PhD

With stories by Dr. Roy Crawford:

"Truth and Tales from Good Old Uncle Roy"

iUniverse, Inc.
New York Bloomington

Goat Production Manual, Second Edition
A Practical Guide

iUniverse books may be ordered through booksellers or by contacting:

iUniverse
1663 Liberty Drive
Bloomington, IN 47403
www.iuniverse.com
1-800-Authors (1-800-288-4677)

ISBN: 978-1-4502-2620-2 (sc)
ISBN: 978-1-4502-2621-9 (ebook)

Printed in the United States of America

iUniverse rev. date: 05/14/2010

This book is dedicated to all my friends (clients)
who taught me more than any textbook could A special thank-you to Jan Schmitz who edited the first and second edition. Her skills with writing and the computer kept me on track.

Contents

List of Tables

Cover Story

The Blucher Grey

This manual is dedicated to the Blucher Grey, pioneer goat of the Canadian prairies: robust, rugged, and reliable. Dr. Roy Crawford, Professor Emeritus in Agriculture at the University of Saskatchewan, developed this hardy breed. As a geneticist, he believes in preserving the genetic diversity of a species. In 1980, the foundation sire was Father Casey, purchased for seventy-five cents (not scents) at the local auction market. Father Casey was selected as a prime example of a hardy indigenous grey prairie goat. The two foundation does were the highly elite Toggenburgs, Mrs. Ross and Mrs. Merrilees. The primary trait selected for was the grey color. Secondary traits were minimum input with adequate output (almost a feral state). To introduce suitable new genetics, Toggenburg billies were used. Now the breed is waiting to be placed on the endangered species list, as only nine does and two billies are known to exist in their native municipality of Blucher. Many offspring have been scattered throughout Canada through the auction market network. Some are gainfully employed clearing brush along the South Saskatchewan river,

Good old Winifred, Fred for short, epitomizes the heart of the Blucher Grey. She milked for four years without the benefit of annual kids, half her udder and being lost in a field of canola for four days.

The cover of this manual features Carolina Brown and her kids (one a true Blucher Grey). Being polled and brown, Carolina was rejected by the breed standards. Although not selected for, the true Blucher Grey has horns; not having horns left the brown Carolina vulnerable to attack. Out of kindness and a sense of a good deal, Dr. Crawford offered to trade Carolina Brown to us for our last border collie pup, Mary, who had flunked all her puppy tests. A fair trade, we thought. Carolina Brown's sire was Pastor Jacob Brown, but her dam is not remembered—obviously she was not an outstanding member of the breed.

To keep the memory of the Blucher Grey alive, Dr. Roy Crawford, alias Uncle Roy, has given me snippets of his life with the Blucher Grey in "Truth and Tales from Good Old Uncle Roy."

My Goats

This guide is a collection of information based on my experiences as a veterinarian and a producer interested in small ruminants. Rather than a how-to guide, this manual is intended to present concepts that, with experience; you can build upon to improve your management skills.

My first introduction to goats was shortly after my family moved from the city "stacks" to the "sticks," as my urban friends would say. I was seven years old, and the move to so much space and freedom was a wonderful adventure.

My mother was allergic to cow milk, so our first farm animal was a pregnant Toggenburg doe that I named Nanny. What else would an eight-year-old name a goat? Instead of doing domestic chores, my responsibility was to care for Nanny. I milked Nanny twice a day. My mother was disappointed that I did not want to drink the milk; I never did tell her the real reason. I knew what went into the milk: Nanny's foot, a few flies, and some not-so-clean straw.

Our next purchase was a large white Billy goat, Spike. I do not think money changed hands. Spike was old and arthritic, but he was kind and very gentle and served his purpose. Spike's only problem was his smell, which left him somewhat of a social outcast most of the year. I was so excited when Nanny had twin bucks, not really understanding that females were the sex of choice.

My goats taught me many important lessons about life. At a time when there was little encouragement for girls to go into the profession, they strengthened my resolve to become a veterinarian. Nanny, Spike, Molly (the polled hermaphrodite), and friends are only memories now. But my fondness for goats was rekindled with the arrival of Carolina Brown (a brown Blucher Grey).

When I moved to Saskatoon to teach large animal medicine at the Western College of Veterinary Medicine, I soon met sheep and goat producers who called me on a regular basis. Colleen Sawyer, one of the producers and a close friend, convinced me that to really understand small ruminants you must own them. She indicated that there was a real need for white-faced range sheep in Saskatchewan, and Columbia sheep seemed to fit the bill. Colleen accompanied me to North Dakota to make sure I would follow through with her suggestions, a move I will never regret. The sheep gave me first-hand knowledge of the ups and downs of the livestock industry. I learned to appreciate the sleepless nights at lambing time and why my management advice would often fall on willing but deaf ears. I learned that textbooks by "experts" often fell short in practical advice. I always advise my students to listen to the seasoned producer, as "nothing upsets a good theory like a little experience." I caution them that they can walk off the farm after advising the client, but the client's livelihood may depend on that advice.

This manual is designed as a guide, one that is not so expensive that if you had it at the barn and some goats ate it, you would worry about the bailiff coming the next day. I offer brief but practical natural management suggestions that hopefully will save you money and time, while giving you peace of mind. Dispersed throughout are stories from *Truth and Tales from*

Good Old Uncle Roy by Dr. Roy Crawford. For those readers who like to keep a diary in the book, spaces for notes are placed in strategic places. For the metrically challenged I provide both the metric and avoirdupois units.

Meg Smart

Chapter 1

The Versatile Goat

There are more than sixty recognized breeds of goats in the world. Ninety-five percent of goats are raised in developing countries and 5 percent in developed countries. Goats are popular with acreage owners "too proud to own a Jersey, too poor to own a cow." Small, early maturing, and prolific, goats can survive on bushes, trees, desert scrub, and aromatic herbs.

On the down side, goats are vulnerable to predators, have a restricted breeding season in the northern and southern hemispheres, and can be labor-intensive, like any dairy operation. As browsers, goats in large numbers can cause environmental damage; they are used to clear land in some countries. Goats and cats are the only domestic animals that can quickly return to a feral (wild) state.

Goats are a main source of food, milk, hides, and wool. More than four million goats produce 4.5 million tons of milk and 1.2 million tons of meat. Worldwide, people eat seven times more goat meat than beef. In North America, chevon (goat meat) is popular with many ethnic groups in large urban centers. Chevon is a low-fat red meat. A meat patty consisting of 60 percent chevon and 40 percent beef produces a low-fat product with a high cooked yield. Goat milk is often used by children and adults who experience allergies and problems digesting cow's milk. There are no religious taboos associated with goats. Goats are also valued for their mohair and cashmere production.

Table 1: Chevon Nutrition
(Source: 2004 USDA National Nutrient Database for Standard Reference)

100g Cooked	Calories (Kcal)	Fat (g)	Saturated Fat (g)	Cholesterol (mg)	Protein (g)
Chevon	143	3.03	0.93	75	27.10
Beef	208	11.07	4.07	84	26.05
Lamb	290	21.12	9.08	93	23.27
Chicken	165	3.57	1.01	85	31.00
Pork	252	14.28	5.25	96	28.88
Bison	143	2.42	0.91	82	28.40
Venison	191	3.93	1.95	113	36.28

Goat Production

Goats are raised in small herds on farms or acreages, where the novice owner often has little knowledge about behavior, management, or nutrition. These owners often depend on how-to books, old wives tales, neighbors down the road, or experienced producers for information on raising goats.

Dairy Production

In the developed countries, large **commercial dairies** are usually located around urban centers. These large operations can experience problems with management, nutrition, and reproduction similar to those of dairy cattle. Mastitis and maintaining quality milk are common problems.

Fiber Production

Angora goats are used in the production of **mohair** (not to be confused with Angora fiber, which comes from rabbits). Mohair is an easily spun, strong fiber that withstands abrasion and retains dyes readily. Since a goat's coat grows continuously, they are sheared twice a year and can yield three to four kg of mohair. Does are not bred until they reach 30 kilograms(kg) (140 pounds [lbs]) to reduce stress abortions. Castrated bucks (wethers) are kept for fiber production and are castrated at an older age to allow for partial development of their horns. Because mohair can represent up to 80 percent of the production income, aspects other than the quality of a goat's coats can be neglected, such as adequate nutrition or reproductive management. However, with better management and nutrition, mohair production can be as high as seven kg/head, and the kidding rate can be greater than 100 percent.

Cashmere is not associated with one breed but often originates from the feral breeds (brush goats). These have a double coat consisting of thick guard hairs with a soft undercoat. The colors of the undercoat are brown, grey, black, or white (the highest monetary value). The soft undercoat is used to produce luxurious cashmere. This coat grows seasonally, and the shed is groomed out and yields on average 80 to 100 grams (g) of cashmere. The kids are often used for meat.

Many options are available to a producer

- National and international sales of quality breeding stock
- Sale of raw fleece to a processor or cottage industry
- Sale of processed fleece as raw fiber through finished products

Meat Production

Boer goats were brought to North America for **meat production**. Embryos were imported and transplanted into locally purchased does. Marketing and close proximity to an ethnic market are major factors in determining profits. The supply of kids is seasonal, and cashmere production is a side venture.

Other breeds include Kiko, Spanish, and Tennessee meat goats. In North America most male kids are sent for slaughter.

Options available to producers

- National and international sales of breeding stock
- Farm gate meat sales
- Commercial live goat sales
- Value-added products

FACILITIES

Goat Behavior

When planning or altering facilities, you must take into consideration the behavior of goats.

Goats living as a group have a definite hierarchal structure. Social dominance is established by head butting; thus, you either should have all your goats dehorned or leave the horns on them all. Horns are a handy set of portable tools that can open gates and smash into areas that appeal to the goat.

Goats are curious and like affection. They are quickly conditioned and prefer to maintain a set routine. Avoid the temptation of handling them like sheep. If threatened or frightened they may bolt in any direction, stand and face the threat, or go into a catatonic state (fainting goat). Their amazing agility often leads them into mischief. They can wreck fences by standing on their back feet and leaning on the fence. As great escape artists, they can clear a four-foot fence or jump down from great heights if the landing is soft. They find new flower buds tasty and like to harvest gardens—often your neighbors'! They are excellent climbers and like to play "kid on the mountain." If no mountains are available, your brand-new, expensive car or that of your neighbor's or visitor's will do.

Goats dislike rain, water puddles, and mud. Their comfort zone is between 55 degrees F (21 degrees C) and 70 degrees F (30 degrees C). Both high humidity and drafts cause stress. As a non-sweating animal, they are less sensitive to declining than rising temperatures. In the summers they need shaded areas to keep cool. Facilities should be flexible and readily adaptable to handle the changes associated with the production cycle or when you retire to a life of leisure.

Shelters

Ventilation is critical, no matter what type of shelter is available. Ammonia build-up is common in poorly ventilated facilities. If you have a problem with coughing in a group or pneumonia in kids housed in a barn, get down at their level and smell the air. Closed wall buildings are difficult to ventilate, and a specialist should be consulted when considering a new facility or renovations to an existing facility.

A problem associated with the cold weather is adequate ventilation in your barn to prevent ammonia, hydrogen sulphide, and methane from building up in the air. The airflow should be such that at the level of the goats there are no drafts but the airflow above the goats is adequate to remove these irritating and toxic gases. This is a definite challenge in western Canada when the temperature dips below -20 degrees C. If you are having problems with air quality or are planning a new facility, I would advise you to contact an agricultural engineer who is familiar with ventilation and barn design. Poor air quality is irritating to the sensitive tissues lining the respiratory tract and can led to a severe outbreak of pneumonia in the does or kids that is not very responsive to antibiotics.

The main shelter should be open, so that as the year progresses the inside can be readily changed with portable pens and barriers. These should be easily taken down, cleaned, and stored. The empty space can also then be easily cleaned. Space requirements should be generous. Allow for 1.4 to 2 feet (42 cm to 60 cm) of feeder space per goat. Allow for 30 to

100 feet (0.9 m to 3 m) per goat in the dry lot (confined to the corrals all year round).

Shelters should open to the south or southeast. The land should slope away from the buildings. Water bowls should be located in higher spots. Place plenty of bedding over a manure pack (manure heats and produces warmth). Place the feed bunk outside to allow for plenty of exercise. Fence-line feeders should be designed so that they cannot be contaminated by urine and feces. A rain shelter over the feeders should be considered.

Stalls should be designed with adequate ventilation. Do not make stalls with solid walls, as this design leads to poor air circulation. Provide an exercise area.

They Hate Rain

"They bow their heads in distress and run for shelter. Including tough old Father Casey, who spent his summers tethered by a long chain near the steel Quonset. He was usually content. But when it rained, he cried like a baby, and all his macho glory and arrogance shrivelled to a bedraggled huddle of misery. Sometimes I felt sorry for him and led him to shelter in the barn. He would trot meekly beside me, murmuring his gratitude ... and then nail me just inside the barn door."

Truth and Tales from Good Old Uncle Roy

Pasture Fencing

Fencing should be designed to keep goats in and dogs (and other predators) out. An example is: 7-foot-long posts spaced at 12-foot centers (2.5-meter posts at 4-meter centers), where fencing is 4 feet high (1.2 m) woven wire with an electric wire 12 inches (30 cm) above the woven fence. With a complete electrical fence the configuration can vary, and portability allows for rotational grazing.

Fence Them In

"If you can. They will go under, or over, or through, and if there is no weak spot they will create one, all with exuberant bounding glee. But an electric fence will stop them. One little wire at shoulder height and one below the knee will do the trick. I labored long and hard at installation, with the whole herd as a curious and interested audience. Then I threw the switch. They clustered in pasture center, staring in dismay first at the fence and then at me. Dismay turned to anger and then to outrage. They hated me, one and all, until winter shut the system down. But they never got out ... except when the power failed."

Truth and Tales from Good Old Uncle Roy

Tethered

Caution: do not use this form of control near your favorite vegetable or flower garden

On small holdings, goats kept as organic lawn mowers are often tied to a movable stake or weights (cement blocks, old car, and truck or tractor tires). Weights allow for some degree of movement and their positions can be changed daily.

Water

Although goats are desert animals and can adapt to periods of low water intake, the modern goat pushed for production requires fresh water daily. A mature goat can drink five liters of water per day. During lactation the water lost in her milk must be added to the five liters. To make the calculations simple, just add the equivalent of her daily milk production. A sure sign of water deprivation is a sudden drop in the group's feed intake.

Records

Two types of records are important.

Barn Records

These are used to keep track of day-to-day occurrences on the farm. Each goat is individually identified, and significant events are recorded throughout her production cycle. These should document her until she leaves the herd. Good records lead to sound management decisions concerning culling, feeding, and herd health. When you purchase a goat, insist on looking at her record and the herd records. Computers are great for organizing data, but for people like me who can do strange things with computers, hard copies of records should be kept. Never keep your only copies in the barn, as goats find them a tasty snack after a bland breakfast of hay.

Records of Performance

These are programs operated by the Canadian and American Goat Society and cover a wide range of production traits. One example of this is the Milk Testing Program.

Milk Testing Program

- **Official 305-day Test**: Official tests supply the owner valuable information on a doe's productivity compared to all goats tested.
- **Five plus Five Test:** Five of the ten milk tests are done by an official paid tester alternated with five herd owner tests.
- **Star One-day Test:** A doe is given a score based on age, stage of lactation, amount of milk produced in a twenty-four-hour period and the amount of butter fat. Her results are compared to does across Canada.

Goat's Milk

The goal is to produce quality milk, which is safe to drink, free from disease-producing bacteria, antibiotics, herbicides, pesticides, and other toxins. The ideal flavor is sweet or salty but free from strong flavors and odors.

Characteristics of Goat's Milk

- High digestibility
- An alkaline pH and a large buffering capacity (neutralizes acids)
- Small fat globules, primarily short-chained fatty acids
- Lacks Alpha-2-Casein (the milk allergen found in cow's milk)
- Naturally homogenized
- Amino acid content similar to human milk
- Deficient in folic acid and B-12
- Low in lactose

Common Questions Asked About Milk

a. What causes blood in the milk?

During the first few weeks after freshening, blood is often seen in the bottom of the container after the milk is cooled. This can be caused by:

- Mastitis (udder usually swollen, either hot or cold to the touch, doe is sick)
- Rupture of small blood vessels (could possibly be genetic)
- Trauma such as vigorous bunting by the kids

Treatment

Separate the doe from her kids except for timed nursing. This may not help, as the kids will be more vigorous, and you may have to bottle or pail feed the kids.

Oral vitamin C or calcium has been tried, with anecdotal evidence of a positive response.

Consult your veterinarian if you suspect mastitis.

b. What causes off flavor?

This can be caused by:

- Being too close to the buck
- Eating weeds/seeds and/or fresh alfalfa
- Exposure to sunlight or UV light (oxidation causes a "cardboard" like taste)
- "Goaty" agitation of fresh milk; mixing warm with cold milk; near the end of lactation (oxidation)
- Possible breed traits (inherited in some families)
- Barn and chemical odors
- Genetic selection by cheese industry in Scandinavia
- Certain diseases e. g., ketosis, mastitis

Treatment

Oral vitamin E may help reduce oxidation and increase the shelf life of the milk (vitamin E can be purchased from a health food or drug store).If possible, remove the doe from the offending diet or source of odors. Heat treat the milk to 57 degrees C (135 degrees F) immediately after milking to inactivate the lipase enzyme, which breaks down the fat causing the "goaty" taste.

c. How do I handle my goat's milk so that it is safe to drink?

To avoid problems with taste, cool milk to 40 degrees F within two hours of milking.

Pasteurization inactivates microorganisms and enzymes that cause the milk to go bad.

A somatic cell count is a measure of the number of white blood cells in the milk as a response to stress or infection (mastitis). The maximum limit for safe milk is 1 to 1.5 million cells per cc (ml). Since this method is used for cow's milk, the counter may not be set for goat's milk, resulting in errors. If your report does not appear correct, check this out. The California Mastitis Test is a barn test available from your veterinarian that screens milk for increased white blood cells. This is more useful in ruling out mastitis than diagnosing it.

Types of Pasteurization

- Heat milk to 145 degrees F (63 degrees C) and hold at that temperature for thirty minutes
- Heat milk to161 degrees F (72 degrees C) and hold for fifteen seconds
- Heat milk to181 degrees F (89 degrees C)and hold for one second
- Ultra high heat milk to 280 degrees F (108 degrees C) to 300 degrees F (116 degrees C) requires special packaging
- Heat colostrum or milk to 131 degrees F (56 degrees C) and hold for sixty minutes for control of caprine arthritis and encephalitis (CAE)

Freeze Storage of Goat Milk

"For years and years the annual cycle at my place has been this: The milk pours out all spring, summer, and fall, and I live in luxury. Then they dry up for a little rest before giving birth on the coldest night of the winter. They nurse the kids for the rest of the winter, giving me some milk in spring. Meanwhile, I have endured many weeks of agony from lactose intolerance of store milk or the displeasure of consuming no milk at all. Then I got a bright idea. Why not freeze it in times of abundance for use in the winter months? I asked many people about a method and they all said "Just freeze it". So I froze it ... seventy liters! I had to buy another freezer to store it all. Was I ever smug. Alas, on thawing, it separated as curds and whey, and no amount of shaking and stirring would bring it back. I couldn't stomach it. But the cats loved it, all seventy liters. What did I do wrong? If you know please let me know and I will try again. Anything to avoid a winter time urpy tum".

Truth and Tales from Good Old Uncle Roy

Often requested Blood Tests

Blood analyses are done in goats for several reasons.

- To ensure raw milk is safe to drink
  ***Note:** always pasteurize milk, to be on the safe side.
- To identify and control certain diseases
- To satisfy government regulations if the goats are exported/imported

The following is a list of tests that could be requested. Before taking the samples, your veterinarian should contact the Canadian Food Inspection Agency (CFIA) of Agriculture Canada for current requirements. If required, a CFIA-certified veterinarian must take the samples and submit them. In the USA, producers are advised to contact United States Department of Agriculture, APHIS Veterinary Services. Specific diseases are covered in more detail in Chapter 5 of this manual.

Toxoplasmosis + Human risk

This is caused by a protozoan and is one reason for pasteurizing milk. Kittens can carry the disease and transmit it in their feces. In goats, it can cause encephalomyelitis (inflammation of the brain tissue), respiratory disease (pneumonia), and abortion. It can infect humans through milk or from handling aborted feti and placenta.

Tuberculosis + Human risk

This is very rare in goats and can be transmitted through milk to humans. The most common test is a skin test. A small amount of antigen is injected intradermal into the skin. After seven days the area is checked for a reaction, which resembles a mosquito bite.

Brucella + Human risk

Although rare, *Brucella melitensis* has been reported in North America and can be transmitted in the milk to humans.

Q Fever + Human risk

This is caused by a rickettsial organism, *Coxiella burnetti*, and is carried by several species of ticks. Shed in milk and transmitted to humans, it can cause anything from mild flu-like symptoms to severe health problems, like endocarditis, an inflammation of the heart muscle.

Blue Tongue (Import/Export)

This is a virus transmitted by a small gnat or sand fly. If importing goats from the USA, this is one of the tests required.

Caprine Arthritis and Encephalitis (CAE) (Disease control)

This is a common viral disease that causes progressive partial or complete paralysis in kids and chronic arthritis, pneumonia, and mastitis (hard udder) in adults. The incidence of the virus in the USA is around 80 percent. The virus is transmitted through colostrum and milk from infected does. To eliminate the virus from an infected herd requires removing the kids at birth and hand-raising them on pasteurized goat's milk, cow's milk, or milk replacer. Ideally, purchase goats from disease-free herds.

Johne's Disease (Disease Control) + Potential human risk

This is a chronic wasting disease caused by the bacteria *Mycobacterium paratuberculosis*. Kids become infected before three months of age by eating the contaminated feces of their dam or other carrier goats. Cattle and sheep can be a source of infection. May be associated with Crone's disease.

Chapter 2

Nutrition

Introduction

The principles presented in this chapter can be applied to all types of goat production. Special nutritional requirements will be indicated where applicable.

Ignorance of goat nutrition is prevalent among small acreage owners and in some larger operations. For example, a client asked me about her goats' nutrition. She had read in her feeding guide that goats will do well on rolled oats. She wondered if she could purchase oats somewhere cheaper than the supermarket. Another fallacy is that goats will eat anything. In fact, goats are fastidious eaters and avoid feed that has been defecated or urinated on. This is why, to avoid waste (of both kinds), they should be fed from elevated mangers with key-hole openings.

As the following table illustrates, although the doe has a higher feed intake than a cow, she produces a higher percentage of milk in relation to her body weight.

Table 2: A Comparison between Nutrient Intake of a Dairy Cow and a Dairy Goat

	Doe	**Cow**
Average Weight (Kg)	60	600
Average 305-day Milk Production (Kg)	1,000	6,000
Milk Production as a percent of Body Weight	17	10
Daily Dry Matter Intake (percent BW)	4 to 7	2.5 to 3.5

A Goat Story

Our family farm was beside a small rural church. One Sunday, my parents felt that the sermon was somehow directed at them, as the minister kept looking in their direction. His theme was "responsible neighbors." I had neglected to tell my parents that I had left the gate open, and the goats had spent a delightful night dining on the minister's vegetables and flowers.

A Goat is a Ruminant

Goats are energetic, inquisitive, and versatile when it comes to food. Goats can survive in a wide range of environments. Their diet consists of a diverse range of feeds. They are anatomically agile (they stand on hind legs) and have a highly mobile upper lip. For pastured goats, browse comprises the major portion of their diet, especially when pastures are mature. Goats are highly selective and eat only plant parts that are palatable to them. They have a high tolerance for bitter-tasting feeds. They can change their diet with the seasons. As a result of their eating patterns, they are used to clear land. In large numbers they can be very destructive to the environment.

The rumen (the first stomach) contains billions of microorganisms (bacteria, protozoa, and fungi) that work together to break down cellulose (fiber) in forages to provide nutrients for their own growth. By-products from their digestion are utilized by the goat. When the microorganisms pass into the small intestines, they are digested and provide nutrients (protein, energy, and minerals) to the host goat. A balanced rumen environment is critical to the health of the goat. Any alteration in the rumen environment can impact on her health and productivity.

When a goat is in a fully alert state she will not ruminate. Rumination[1] (cud chewing) is elicited by light, so cud chewing may decrease in the winter in a dark barn with minimal light. More light at timed intervals may help production.

Feed Restrictions

Adult goats (not pregnant or lactating) can adapt to medium and long-term malnutrition. The liver produces energy by using muscle protein, which causes loss of muscle mass. This is no way to treat a friend and provider.

Is the rumen really necessary?

The short answer is yes. The rumen allows animals to consume plant material and retain it long enough to allow the bacteria, protozoa, and fungi to penetrate and digest it. Through the microorganisms in the rumen, all ruminants become capable of utilizing cellulose and other plant material, but large variability exists between ruminants regarding the quality of feed that each is capable of digesting. Most ruminants develop a preference for feeds with a high satiety level (fills them up). Through the evolutionary process, three types of ruminants have developed.

Concentrate selectors

They cannot tolerate large amounts of fiber in the diet. Examples are moose and white tail or mule deer.

Intermediates

They are adapted to browsing and grazing. They prefer young and immature grasses. As the grasses mature they switch to browsing. (See Pasture and Browse Management later in this chapter.) Examples are goats and elk.

Forage selectors

They maximize the extraction of nutrients from fibrous plant material, which results in a slower rate of passage through the digestive system. Examples are sheep and cattle.

The physical size of the goat and rumen, space in the abdomen, forage quality, and water intake are key factors in the control of feed intake. Poor quality forages take longer to digest and remain in the digestive tract longer. Inadequate water intake will limit feed intake. Developing fetuses will limit space in the abdomen. A growing uterus and a large amount of fat in the abdomen can limit the size of the rumen and the amount of feed it can hold and digest. This can restrict the nutrients available to the doe and her fetus and predispose her to metabolic problems, such as pregnancy toxicity.

Nutrient Requirements

Most nutrient requirements are based on an interpretation of research published in the scientific literature (see Tables 8, 9, 10, 11, 12). Most of the experiments are done on a limited number of animals maintained in a comfortable working environment (not 40 below zero). Most practicing nutritionists use these requirements as the first step in ration formulation and then add adjustment factors to account for environmental temperatures, type of shelter, level of stress, and stage of production.

Feed testing

Testing feed is a valuable management tool if the samples are taken properly and accurately analyzed. **Note**: See your local feed store or agricultural representative for correct sampling procedures.

Forage quality is the key factor in obtaining optimum milk production over a doe's lactation. Because of lack of knowledge, economics, and small numbers, most goat producers do not have their feeds tested. When a producer must buy feed, either feed testing prior to purchase or purchasing tested feeds is a wise economic decision. This is important in times of feed shortages when there may be little difference in price between poor and good quality feeds.

To accurately test forages you must take composite core samples[2] from at least twenty small square bales or five large round bales. The bales should represent the forages to be fed over the feeding period. A sample of grain representing the type fed should be submitted. A forage package analysis usually includes moisture, protein, calcium, phosphorous, potassium, sodium, nitrates, and energy estimation. Grains need to be tested for moisture, protein, and energy. A more detailed analysis is dependent on the conditions where the feeds were grown. Trace mineral analyses are an added cost.

Why should I have my feeds tested each year when they are harvested from the same fields?
Environmental conditions are seldom the same from year to year. The quality of the feed can be altered by weather conditions while growing and during harvesting. Some years the protein in the grains is higher than others. Hay from drought-stressed or frozen cereal crops can be high in nitrates.

When I receive the feed test results, they mean nothing to me. How do I use them?
To help you, I will take you step by step through formulating a diet for your does in late pregnancy. How you formulate your does' diet depends on the feedstuffs available and affordable.

Pasture and Browse Management[3]
Targeted Grazing[4]

Targeted grazing is a means of using livestock to manage and sustain an ecosystem. In Canada the management of shrubs as browse for goats may be considered a novel concept. Goats are opportunistic feeders who can shift their diet selection among browse (shrubs and trees), forbs (broadleaf weeds and legumes), and grasses. Their small muzzle, narrow tongue, and ability to stand on their hind legs allow them to access the leaves of shrub and trees. Animal preference, palatability, and digestibility of browse are related to the amount of secondary plant compounds (chemicals) present.

Tannins are the most common of these compounds. Tannins are of two types: soluble and condensed. Soluble tannins can have a toxicological effect when absorbed into the goats' blood stream. Within the digestive tract, condensed tannins can bind with the microbial digestive enzymes, making them ineffective. This can lead to a protein and energy deficiency. The recommendation regarding tannin content is 2 percent of the total diet. Goats appear to tolerate a higher concentration of tannins than cattle. The tannin content is highest when the shrub is rapidly growing. The rumen microorganisms can adapt to help counteract the negative effects of tannins.

Browse plants peak nutritionally in the spring. The magnitude of this peak is dependent on the winter moisture. Little moisture can result in new twigs and leaves not developing, lowering the nutrient content (though the degree to which this occurs varies with the species of the shrubs). During a drought shrubs store nutrients in both the stem and roots, thus the nutritional value of shrubs is maintained during drought better than grasses, but crude protein will still decline.

Limited research has been done on the mineral content of browse. What has been done indicates that they are a little higher than the forages but still mirror that of grasses. Thus supplementation is recommended.

Shrubs and trees should be considered in the overall management of pastures for goats. They cannot be relied on to provide supplemental nutrition if the pastures are mature or over grazed, but they can add variety and nutritional support if managed along with the pastures.

Voluntary Versus Dry Matter Intake

Before I go any further and confuse you any more, I must define two factors that are difficult to measure but essential to understand when you evaluate or formulate rations. They are voluntary intake (VI) and dry matter intake (DMI).

Voluntary Intake (VI)

This is the amount of a single feed that a goat will eat in a day under ideal environmental conditions (individually penned, a comfortable ambient temperature, no mental or physical stress). Thus VI is a measure of the quality and palatability of the feedstuffs. Normal voluntary intake is altered by gut fill and environmental temperature. If cold-stressed, the goat will eat more than her VI in an attempt to meet her energy requirements. In late pregnancy her VI is limited by the growing fetuses, which reduces the amount of food that can be digested in the rumen and intestinal tract. Thus, in order to meet her requirements for late pregnancy, the nutrients must be concentrated in a smaller but palatable package. If this is not done, she will be at risk for pregnancy toxemia and/or low blood calcium (hypocalcaemia).

Weathered, moldy, or unpalatable feed will reduce VI, but you must determine by how much.

In late Pregnancy with twins, VI is reduced 15 percent; with triplets VI is reduced 20 percent.

Table 3: The Suitability of Forage and Pasture Quality *

Factor	Crude protein percent	Total Digestible Nutrients (TDN) percent	Voluntary intake as a percent of body weight Suitability for goats	Type of forage and stage of pasture growth
Poor Quality Forage	<5	<50	.5 to 1 percent Not suitable If fed, feed only 10 percent of the total ration Will need additional better-quality forages, protein supplements, and grains to make up the nutrient deficit	Mature grasses Green feeds, straw, and mature grain Full bloom alfalfa Straw Mature pastures past full flower In a stubble field there may be enough mature grain, but a protein supplement will be needed
Medium Quality Forages	5 to 10	50 to 55	1.5 to 2 percent Not suitable If fed, feed about 10 to 15 percent of the total ration Energy protein and other nutrient deficits must be addressed	Cereal hays and crops: soft dough stage Grasses, full bloom Legumes >75 percent bloom Pastures, full flower
Good Quality Forages	11 to 14	55 to 58	3 percent Suitable Energy supplements in the form of grains may be necessary during the final finishing stage for meat and in late pregnancy	Cereal crops and hay milk, early dough stage Grasses, early flower Legumes, 10 to 20 percent bloom Pastures, early flower
Excellent Quality Forages	>14	>58	>3 percent Suitable Depending on economics and the availability of feed, may limit the amount fed and balance energy with grains	Grasses and cereal crops, no flower, early growth Legumes <10 percent bloom Pastures, early growth, first 10 to 20 cm

* Note: Grasses and cereal crops can be deficient in calcium (Ca) and phosphorous (P) and will need a 1:1 Ca: P mineral; legumes can be low in P and high in Ca and may need a 0:1 Ca: P mineral. All will require a salt and appropriate trace mineral supplement.

Dry Matter Intake (DMI)

This is a measure of maximum feed intake that a goat can physically eat during the day when all the water has been removed. This amount of feed must provide all the goats' daily nutrient requirements. Normal dry matter intakes for each stage of production are the key to successful nutritional management. For a feedstuff to be included in a ration formulation, the nutritional parameters must be based on the removal of the water present in the original feedstuff.

- Early pregnancy DMI 3.0 percent BW
- Mid pregnancy DMI 3.0 percent BW
- Late Pregnancy DMI 3.0 percent BW
- Early lactation DMI 4.0 to 7.0 percent BW*
- Mid Lactation DMI 3.0 to 4.0 percent BW*

  * Related to milk production

How do I use these concepts to formulate a ration?

Caution! These are not concepts you should tackle while under stress (i.e., lack of sleep at kidding time). Try only when your brain is working at full capacity.
Example:
I will use a 65-kg doe in late pregnancy as an example. Her crude protein requirement is 12.0 percent of DM, and TDN[5] requirement is 55 percent to 60 percent of DM. She is pregnant with twins (at least she had twins last year), and she is in acceptable body condition. All her nutrient requirements must be packed in a DMI of 3.0 percent of her body weight.

If fed **poor quality forages,** her voluntary intake is 1.0 percent BW. To meet her DMI and Total Digestible Nutrients (TDN requirements she must be fed a protein (e.g., canola meal) and energy (e.g., barley) supplement in the remaining 2 percent DM. To meet her DMI requirements without supplementation with this particular forage may lead to metabolic problems. This is not the forage to feed, even if it is free.

Next, let us consider **medium quality forages.** Her voluntary intake is 1.7 percent BW. To meet her DMI she must eat an additional 1.3 percent DM in the form of a protein and perhaps an energy supplement. This you can do as long as you gradually increase the supplement. You could purchase this hay, but being an excellent manager you decide to look for better, since this hay is weathered and moldy (low palatability).

You have finally found **a good quality hay**. Her voluntary intake will be 3.0 percent, thus she may not need additional supplements.

These same principles can also be applied to pastures, as to whether supplementation will be required.

By doing these calculations at a time when hay becomes available, long before kidding, you can save money and prevent panic buying of hay. If you put up your own hay, this will encourage you to put it up in good condition, if possible, or at least you can tailor your feeding program to minimize the costs. Last-minute preparations can be costly where nutrition is concerned.

Key Ingredients in a Ration

Forages

Forages are the key to maintaining a healthy rumen and to successful ration formulation. Good quality forages are essential but not always available. A goat is physically limited in its ability to eat enough poor quality forge to meet its requirements. This is particularly critical in the last third of gestation. **Goats will waste 10 to 20 percent of poor quality forages**. These types of forages will require a protein and/or grain supplement to meet requirements, thus adding to your feed costs. Large round bales lose moisture over the winter and can weigh 10 percent to 15 percent less than when you purchased or started to feed them. Depending on the quality, you may have to adjust intake.

When purchasing feeds, you must factor in wastage, moisture loss, and cost of supplementation to the purchase price to determine if you are getting a fair deal.

Straw

Straw is considered a poor quality forage. Goats do not like it and waste about 20 percent of it. Straw has 3 percent to 5 percent crude protein and 45 percent TDN. A goat's normal voluntary dry matter intake of straw is less than 1 percent of her body weight. Protein and energy supplements will be needed to balance the straw. Some producers grind the straw, hoping the goats will eat more, but you cannot fool a goat. Grinding does not change the quality or digestibility of the straw. If forced because of environmental or metabolic stressors, a goat will eat more and lose weight, with diminished production. If close to kidding, she will be prone to develop pregnancy toxemia.

Green Feed

This is versatile forage if cut and baled at the right time. Green feed is from a cereal crop (primarily oats) that has been cut as forage before maturity. It may have been a crop that was planted late and froze before it matured. Some farmers plant a cereal crop to harvest as forage or use as a fall pasture. Nitrates can be a problem in drought-stressed or frozen green feeds. **Note: Your water supply can be a significant source of nitrates.** If a cereal crop freezes, cut it for hay immediately or wait between eleven and fifteen days and then cut it. Before cutting, take samples from many areas in the field and have them tested for nitrates. Your veterinarian, agricultural representative, or feed testing lab may offer the rapid test for nitrates. If the green feed has a nitrate concentration of over 0.5 percent, then the feed should be diluted with another type of forage. The quality of green feed varies from poor to good, so a feed analysis is essential, especially if you are buying it. A good quality green feed is palatable and adequate in protein and energy but low in vitamin A, vitamin E, and all the minerals.

Grains

Grains are essential when the diet contains poor quality forages. They are fed to increase the energy content of the diet, particularly in late pregnancy and during lactation. Grains are fed whole. There is no advantage in processing them, as whole grains are well digested. The extra cost of chopping is not necessary. Barley is the most commonly fed grain. Oats can be fed, but some goats tend to choke on whole oats, though they seem to cough them up with no apparent after effects. **Limit oat intake to 25 percent of diet DM because of high fiber content.**

Vitamins, Minerals, and Salt

Vitamins, minerals, and salt are fed to balance the salt and mineral portion of the diet. To ensure adequate intake you should always feed these products loose, not in block form. If your goats refuse to eat a mineral, try a different manufacturer, as each company has their own special formulation.

The most common formulation of goat mineral is two parts calcium to one part phosphorous. This is not the best if alfalfa (high in Ca) is fed. In some areas added copper is required if the feeds are low: < 5 mg Cu/kg diet dry matter (DM); if the water is high in sulphates (Table 3); or the feed is high in molybdenum (> 10 mg/kg DM). Selenium is added to most minerals, and you must not feed your goat Se from multiple sources. A loose cobalt/iodized salt should be available to the goat either alone or mixed with the mineral (one part salt to three parts mineral). In the winter, and for totally confined goats, the mineral should be fortified with vitamins A, D, and E. **Note: Mineral manufactured six months prior to feeding may not have adequate levels of vitamins A, D, or E.** Ask your feed salesperson the date that the mineral was manufactured.

Did you know? Goats are not as prone to copper toxicity as sheep. Thus cattle mineral can be fed to goats. The A, D, E injectable vitamin does not contain adequate vitamin E. Vitamin E is added to preserve the other two vitamins.

Water

Water, a key ingredient, is often overlooked. Through evolution the goat has adapted to limited water intake and short-term shortages in response to desert conditions. At high environmental temperatures (>38 degrees C, 99 degrees F), goats conserve water by decreasing the water in urine and feces. The goat is close to the camel in its ability to conserve water. At some times of the year goats will obtain enough water from plants. Water should not be overlooked as a source of macro and micro minerals.

Table 4: A Summary of Acceptable Water Quality for Livestock

Element	Maximum Safe Level (mg/l)	Comments
Hardness	180	Very hard, sum of calcium and magnesium in the water Soft water can be corrosive and tends to leach copper (Cu), zinc (Zn), cadmium (Cd) from pipes, and carry toxic elements from pipes or soil
Total Dissolved Solids (TDS)	7,000 to 10,000	The impact on production depends on mineral composition
Magnesium	800	May be laxative
Sodium	800	Laxative and increases thirst and urination
Calcium	1,000 to 2,000	Can contribute to milk fever in dairy cows
Iron	175 to 300	> 1,500 can reduce dietary copper availability
Sulphates (SO4)	< 900	Reduces copper availability Associated with polioencephalomalacia (thiamine deficiency)
Nitrates (NO3)	< 100	Additive with nitrates in the diet Decreases vitamin A, E, iodine, and phosphorous availability (species dependent)
Zinc	25	
Copper	0.5 to 1.0	30 percent of a sheep's daily requirements
Selenium	0.1	Ten-fold margin of safety
Chlorides	1000	> 250 causes corrosion of plumbing equipment
Mercury	0.01	Toxic
Lead	0.1	Toxic
Fluoride	< 3.0	Can cause mild mottling of teeth

Basic Feeding Guidelines

The following sections of the manual are designed to help you evaluate and formulate economical rations. The two most common nutritional problems are under- and over-nutrition. Nutrition is not an exact science. When I investigate a nutritional problem, the solution can be simple or very complex. The body is tolerant of short-term nutritional abuse. Nutritional problems can have a long-term negative impact on your operation and its profitability. Unless you are involved in this venture for fun (you are a lottery winner or independently wealthy), pay close attention to changes in body condition, as it reflects the health and nutrition of your goats. Clinical symptoms occur if the abuse is long-term, extreme, or at a critical production stage. The response of the body varies according to the stressors that influence the nutrient requirements.

There are two critical components in the formulation of an economical ration to optimize production.

- An accurate feed analysis
- A minimum of one accurate weight scale. If you measure feedstuffs in tobacco tins or by handfuls, you don't have to abandon these methods of your forebears; just measure the weight of the contents.

Pregnancy and the Dry Period

The feeding program for your does during the last part of lactation and the dry period can impact overall milk production in their next lactation. From early to late gestation, feed good quality forage. Does should be in good flesh and not too fat. Some grain may be needed near the end of the dry period (the last two to four weeks).

The dry period should be forty to sixty days long to allow the doe to regain body condition if thin, and it gives the mammary glands time to rest and prepare for the next lactation. If a doe is thin going into early lactation, she will have limited energy reserves (primarily fat, some protein) to draw on to support maximum milk production. This may create a domino effect, leading to a poor conception rate and early culling.

During the first two-thirds of her pregnancy, depending on body condition, weather conditions, and the type of shelter, feed a good quality forage, loose cobalt iodized salt, and mineral (composition of mineral is dependent on the type of forages fed).

Table 5: The Changes in Body Reserves and Nutritional Requirements of a Mature Doe through Her Production Cycle

Stage of Production	Nutrient Requirements for …	Body Fat	Body Protein	Energy Balance
Drying off	Maintenance	Stored	Stored	Positive
Last 5 weeks of pregnancy	Maintenance + high fetal demand*	Mobilized	Mobilized	Negative
Over first 2 months of lactation	Maintenance + high lactation demands	Mobilized	Mobilized	Negative
After the first 2 months	Lactation demands stabilized	Stored	Stored	Positive

(*Pregnancy toxemia will develop if stored reserves are inadequate to meet her demands. The fat stores are mobilized first)

Depending on management practices, weather, and when the doe is due, pasture, good hay, or silage can be used as forage. Some nutritionists recommend that good quality alfalfa hay should not be the sole roughage fed because of its high calcium content. If needed, feed a 12 percent protein/grain mix at .22 to .45 kg/head/day. (.5 to 1 lb/head/day)This will lower the calcium content of the diet and conserve alfalfa hay

Over the last third of pregnancy the fetus is growing rapidly. If the doe's diet is inadequate, she will supply the nutrients for fetal growth by breaking down her own tissues. Complications associated with this can include body weight loss and pregnancy toxemia (ketosis). Angora goats require extra nutrients in the last half of pregnancy to prevent stress abortions. They have a high requirement for fleece (fiber) growth at this time. You must ensure intake of at least 16 percent crude protein.

Start her gradually onto concentrate late in gestation (the last two to three weeks). On average she should be receiving 0.45 kg (one pound) of a grain/protein supplement just before kidding. Feed the correct ratio of forage to concentrate (grain + supplemental protein). Avoid going to less than 40 percent forage in the diet. Nutrient composition of the concentrate will depend on nutrients in forage. Ensure an optimum voluntary intake of good quality forage (e.g., alfalfa).

Postpartum and Lactation

A lactating doe can voluntarily eat 6 to 11 percent of her body weight in total dietary dry matter. Concentrate should be increased gradually over a two-week period to 0.9 kgs (2 lb) per day postpartum, or feed a half a kilogram of concentrate/liter of milk/day (0.5 kg/L of milk) (3 lbs/gallon) One publication recommends forage intake should be 1.7 kg of forage DM/100 kg live weight (LW),(4lbs/220 lbs LW) and concentrate should be gradually increased following kidding from 400g (.5 lb)to 500 g (1 lb)of dry matter/100 kg (220 lbs)LW.

As forage quality declines, more concentrate is needed, but always keep in mind the ratio. Don't go under 40 percent roughages. Under conditions where forage availability is limited, this rule may have to be broken temporarily. **Note: Whenever ration changes are made, make the change slowly over time to avoid digestive upsets.**

Kids: One To Seven Days after Birth

This is a critical time. Free choice colostrum (first milk) is essential for many reasons.

- It provides antibodies that protect the kids from diseases that their mother has been exposed to naturally or by vaccination (**Note: To date, commercial cow colostral products only contain antibodies associated with cattle diseases**).
- It contains natural antimicrobial compounds that protect the newborn's gut against harmful organisms, allowing time for the normal microbial flora to develop.
- It contains high concentrations of minerals and vitamins A, D, E, K, and carotene, as the kid is born with minimal reserves of these essential nutrients.
- It contains cells from the mother's blood that will turn on the kid's immune system.
- It contains important growth factors that help stimulate early growth.

An Emergency Substitute Colostrum

820 ml warm cow, goat, or sheep milk
2 ml of corn syrup (not table sugar; corn syrup is more digestible)
5 ml cod liver oil
1 whole egg + shell
Mix in a blender and feed 140 ml warm, four times a day for two days. This provides kids with fat, protein, minerals, vitamins A, D, and E, but no antibodies.

Whether the kids are left to nurse on their mothers until weaning or are started on ***milk replacer*** at three days of age is a management decision based on the goals of the operation. Table 7 highlights the key factors involved with the successful feeding of milk replacers. Milk replacers (ingredients from cow's milk) seldom meet the kid's requirements. If the kids are stressed, health problems related to inadequate nutrient intake can occur.

For the first two to three weeks, feed a high quality milk replacer with all the protein derived from milk sources. If you open the bag and the replacer smells rancid, is yellow to brown in color, and does not remain in suspension when mixed, do not feed it. Return it to the feed store.

a. What is a creep area?

A creep area is a place kids can go but adults cannot. Provide good quality hay and a starter ration for kids to eat without competition with adults. If kidding occurs during cold months, the creep area should have a heat source which will attract the kids. Kids started in the creep area are easier to wean.

b. Is feeding eggs an old wives tale?

I am often called during the spring by producers who have heard that eggs are good for scours and as a supplement. After investigating this question, I found that eggs are unique, the chicken's answer to colostrum. They are highly digestible. If fed whole including the ground-up shell, they are a good source of minerals, protein, fat, and vitamins and provide 6.5 percent of a kid's daily energy requirement.

Homemade Electrolyte for Scours

1 can beef consommé
1 pkg dry Certo (fruit pectin)
1 teaspoon lite salt (potassium chloride [KCl])
2 teaspoons baking soda

- Add water to 2 quarts (distilled if tap water high in nitrates, or if Total Dissolved Solids (TDS) > 800 mg/l).
- Mix well. Divide into 4 equal portions to be fed at 2- to 4-hour intervals.

Kids: Five Days to Weaning

Requirement tables do not include kids less than one month of age; the assumption is that they are nursing.

The following is based on my interpretation of the literature for other ruminants:
Maintenance requirements: 55 Kcal DE (digestible energy)[6]/kg BW

Requirements for gain: 3.0 Kcal DE/g of gain

Results of a trial investigating the effect of frequency of feeding on the growth and health of kids indicated that feeding the entire quota of milk once a day versus twice or three times daily required less labor. The kids fed more than once a day grew faster but the incidence of bloat, indigestion, and death were comparable. (This trial did not increase the total amount of milk fed, only the frequency.)

Table 6 The Importance of Milk Production in Meat Goats

	# of Kids	SINGLE			TWINS			TRIPLETS		
Doe's Milk	(Birth weight 4 kg)	Need for mainten-ance	Kcal left for gain	Gain /kid	Need for main-tenance	Kcal for gain	Gain /kid	Need for main-ten-ance	Kcal for gain	Gain /kid
kg/d	kcal/d	Kcal/d	Kcal/d	g/d	Kcal/d	Kcal/d	g/d	Kcal/d	Kcal/d	g/d
2	1,480	220	1,260	420	440	1,040	173	660	820	91
4	2,960	220	2,740	913	440	2,520	420	660	1,300	256
6	4,440	220	4,220	1,407	440	4,000	667	660	3,780	420
8	5,920	220	5,700	1,900	440	5,460	913	660	5,260	584

The table above illustrates the importance of selecting and feeding for milk production in meat does to optimize weight gain in nursing kids. Recording the kids' weights at weaning and calculating the average daily gains will be an indication of the doe's milk production. Weighing kids on a regular basis while nursing will help you determine when her milk production will no longer support adequate gain. These records are critical when culling does, choosing replacement bucks and does, and buying or selling breeding stock.

Table 7 A Comparison of Goat, Cow, and Sheep Milk with a Commercial Milk Replacer

	Dry Matter percent		Protein percent		Fat per-cent		Lactose percent		Ash per-cent*		Energy Density Kcal/gm	
	90**	As fed	90	As fed	90	As fed	90	As fed	90	As fed	90	As fed
Goat	90	13	26	3.3	29	4	29	4	5	0.8	5.1	0.74
Cow	90	13	23	3.2	26	3.6	34	5	5	0.8	4.9	0.7
Ewe	90	18	26	5.5	35	7	32	4.5	5	0.9	5.3	0.77
Lamb Milk Replacer	90		20		25		41		4		4.8	

* If ash content in the replacer is >1, plant-based protein is present (this is all right for kids over three weeks)

** Comparison is based on the dry matter content of the powder prior to mixing with water

Notes on Feeding Milk Replacers

- Always follow the directions on the bag (often difficult). If confused, mix approximately one part of powder to four parts of water. If your water is high in sulphates and/or iron (see water quality), scours may develop until the kids adapt. If you have access to better quality water, mix the waters and gradually increase the amount of the poor water.
- The instructions on the bag usually suggest feeding twice a day. Do not exceed this amount per meal. If the kids are allowed to gorge themselves twice a day, exceeding the stomach's ability to digest the replacer, the milk replacer may start to ferment in the abomasum and cause bloat.
- Keep in mind that the milk replacer is designed to meet the kid's minimal requirements and does not consider any allowance for stress. Your kids may not grow well until they start eating adequate creep rations and good quality hay.
- If abomasal bloat occurs, one teaspoon of ground ginger in a small amount of warm water may help. The kid can die from bloat. Pressure can be relieved by releasing the gas through a 14-gauge one-inch needle pushed through the abdominal wall to the right of the navel into the abomasum. These problems can be reduced by feeding all-milk protein milk replacers as directed. The replacer should contain a minimum of 20 percent protein and 25 percent fat. If you want to feed more to improve the kids' growth, mix according to directions but feed smaller individual meals four to five times a day. If bloating persists, mix a few drops of formalin into the replacer before feeding, or try another manufacturer of replacers.
- Kids can be fed cow's milk, but generally they do not grow as well because cow's milk is not as digestible.
- Acidified milk replacers are designed to be fed through a pail with several nipples attached. This allows the kids free access to milk through the day, thus simulating a more normal nursing pattern. The milk replacer is designed to be fed cool so as to not ferment. Hygiene of this system is critical.

Kids born in the cold weather are very vulnerable to hypothermia and frostbite. You can treat the hypothermia with heat, intravenous or intra-peritoneal dextrose, and warm colostrums orally. Be aware that frozen ears and limbs may not be evident at the time. The severity of frostbite will become evident several days to weeks after birth. The most vulnerable parts of the body are the ears (unless you raise La Manches) and hooves. The ears may become swollen several days after birth. In a few days to a week the swelling goes away; the frozen portion turns black and eventually falls off. The hooves and legs take a little longer to show the effects of frostbite. The first sign is usually lameness. When you examine it closely the hoof will be cold, and there will be a separation where the hair meets the hoof, with serum oozing from the area. Eventually the hoof will fall off. If the limb is frozen farther up, the leg will be black and the dead tissue will be separating. In either case, the prognosis is poor for the future of the kid. If you decide to raise the kid for slaughter, you must consider whether this is a humane treatment. The affected kids should be separated from the main group and raised in non-competitive environment.

Kids: Weaning

Kids are usually weaned at eight to twelve weeks of age. Weaning is easier if they have access to a **creep ration** of good quality hay, a 16 percent protein grain mix, free choice loose cobalt iodized salt, minerals, and water. This ration should be available to them no later than four days after birth. If coccidiosis is a problem, a coccidiostat may be fed. At weaning they should be eating between 0.22 to 0.45 kg of starter per head/day. In a research trial, kids weaned at 10 kg BW and fed a ration of 18.0 percent crude protein with a hay-to-grain ratio of 60:40 weighed 26 kg at 8 months of age and reached target weight (30 kg) for breeding at 9 months. Later-weaned kids gained faster but cost more to raise.

Kids: Weaning to Breeding

This group requires nutrients for maintenance and growth. The starter ration can be fed at 0.22 to 0.45 kg/head/day. This may be costly if purchased prepared. A home diet of whole grain (oats or barley) and a protein supplement (e.g., canola meal or soybean meal) can be fed, and the composition adjusted for the nutrients in hay (feed testing will pay for itself). Avoid feeding silage and urea.

Table 8: Daily Nutrient Requirements of Goats

Maintenance Requirements	Daily Dry Matter Intake (DMI)		Total Digestible Nutrients (TDN)	Crude Protein	Ca	P
Body Weight (kg)	Percent BW	kg/day	kg/day	g/day	g/day	g/day
10	3	0.3	0.16	22	1.0	0.7
20	2.5	0.5	0.27	38	1.0	0.7
30	2.2	0.7	0.36	51	2.0	1.4
40	2.0	0.8	0.45	63	2.0	1.4
50	2.0	1.0	0.53	75	2.1	1.4
70	2.0	1.4	0.70	96	4.0	2.8
100	2.0	2.0	0.9	130	5.0	3.5
Add for last 3 weeks of pregnancy*		0.7	0.40	82	2.0	1.4
Growth requirements Kg of gain / day/kid						
0.05 kg/day		0.18	0.10	14	1.0	0.7
0.10 kg/day		0.36	0.20	28	1.0	0.7
0.15 kg/day		.54	0.30	42	2	1.4

* Examples:

A 50 kg doe in last 3 weeks of pregnancy requires a DMI of 2 percent BW or 1 kg/d for maintenance. For pregnancy, add .07 kg total DMI 1.7 kg or 3 percent BW.

A 10 kg kid requires a DMI of 0.3 kg For maintenance: for a gain of 0.1 kg/day requires an additional 0.36 kg for a total DMI of 0.66kg/d

Table 9 Additional Requirements for One Kg of Milk Production per Day

Percent Fat	TDN (g)	CP (g)	Ca (g)	P (g)	Vit A (IU)	Vit D (IU)
2.5	333	59	2	1.4	3,800	760
3.0	337	64	"	"	"	"
3.5	342	68	"	"	"	"
4.0	346	72	3	2.1	"	"
4.5	351	77	"	"	"	"
5.0	356	82	"	"	"	"

Table 10: Daily Macro Minerals for Goats and Some Key Points

0Minerals	Requirement (% of DMI)	Common Sources	Comments
Sodium (Na)	.04 to .10	Supplied by free choice loose salt.*	If absent or low in diet goats will crave it. They will chew on anything that is salty.
Chloride (Cl)	——	Present with Na in salt.	
Calcium (Ca)	.21 to .52	Forages high in alfalfa Commercial minerals e.g.,1:1, 2:1/Ca:P Limestone, bone meal, dicalcium phosphate High in some water supplies	If deficient in kids, skeletal deformities, fractures, and lameness can occur—often associated with P, protein, and Vitamin D deficiencies (no sunlight). If deficient in adults, it can cause stiffness and recumbency following stress, e.g., kidding and early lactation.
Phosphorus (P)	.16 to .37	High in grain, and low in forage, especially if poor quality or from areas of deficient soils Commercial supplements	Deficiency in adults can cause fertility problems or osteoporosis. Low first-service conceptions Silent estrus
Magnesium (Mg)	.04 to .08	Commercial supplements Deep well water	Lush pastures can cause a deficiency; goats become stiff and convulse (tetany).
Potassium (K)	0.5	High in good quality forages, e.g., alfalfa.	Requirements can increase up to 0.8 percent in stressed animals, e.g., at kidding and early lactation.
Sulphur (S)	0.14 to 0.26	High in water from deep wells.	If S is high in diet >0.5 percent can interfere with Cu. Can be a risk factor in polioencephalomalacia (thiamine deficiency).

* Salt and mineral should be fed in loose form as small ruminants may not be able to lick enough

Table11: Trace Mineral Requirements of Goats ***Toxic levels

Minerals	Requirements (mg/kg DM)	Common Sources	Comments
Copper (Cu)	5 [> 25***]	Most feedstuffs in western Canada contain less than 7 mg/kg or parts per million (ppm), especially straw and grain.	Deficiency in newborn causes lack of coordination and recumbency related to impaired brain development. Can delay onset of nervous signs until 1 to 2 weeks of age.
Zinc (Zn)	35 to 50 [>1,000]	Most diets are adequate.	Deficiency causes crusty skin lesions and hair loss.
Iodine (I)	0.1 to 0.8 [>50]	Cobalt iodized loose salt Cannot lick enough block salt	Deficiency goiter in newborn, or hairless, stillborn, or weak
Selenium (Se)	0.1 [>2]	Most commercial minerals are fortified. Deficient in feeds grown in the gray-wooded soils. Injectable products Passes to the nursing kid through milk	Deficiency causes white muscle disease (WMD) in kids. Must avoid an excess of Se Vitamin E deficiency can cause WMD independent of Se.
Manganese (Mn)	20 to 40 [>1,000]	Low in straw and grains	Deficiency in the fetus results in skeletal defects and contracted tendons.
Iron (Fe)	30 - 50[>500]	High in soil and some water supplies Deficiency is not usually a problem unless secondary to a chronic blood loss.	Deficiency causes anemia and weight loss. Milk low in iron Secondary deficiencies occur with chronic blood loss, e.g., blood sucking internal and external parasites.

Table 12: Vitamin Requirements of Goats

Vitamin	Requirements	Key Functions	Conditions Associated with a Deficiency	Key Symptoms of a Deficiency & Treatments
A	1,000 to 9,000 IU/day Varies with production groups	Normal vision Role in reproduction Maintains healthy skin Promotes resistance to infections	Extended dry periods where the supply of green forages is limited Weathered hay High levels of dietary nitrates Low dietary zinc	Loss of night vision to blindness Decreased fertility in buck and doe Increased risk of infection High neonatal mortality
D	100 to 500 IU/day	Absorption of Ca & P	Little or no access to sunlight Winter months (Oct to Mar)	Primarily in the growing kid—lameness, skeletal abnormalities, and fractured bones
E	15 to 40 IU/day 500 IU / day improves milk and carcass quality.	Maintains healthy cells by neutralizing harmful products of cell metabolism Improves shelf life and flavor of milk	Feeding weathered/poor quality hays Severe stress can precipitate, e.g., transportation	White muscle disease in kids Can use Vitamin E capsules
B Mainly Thiamine	Unknown, as most are produced in a healthy rumen.	Important in brain metabolism	Digestive upsets involving the rumen Feeding amprolium over a long term Water high in sulphates Koshia weed	Convulsions, blindness Responds to thiamine
C	Unknown	Maintains integrity of blood vessels	Unknown	Blood in the milk—this is only speculation

Winter Considerations

As winter approaches, the local coffee shop patrons and grocery aisle visitors will stop to speculate: “What will winter be like this year?” What can we expect; after all, we live in the cold Canadian West? Farming is a hazardous venture at best, and we are at our best when the worst occurs. Once the worst is over, we can relax and laugh at how we conquered the winter again.

Marion (Meg) Smart DVM, PhD

The following are helpful hints on how to deal with the bitter cold of winter. But first, you might want to remember this advice from Mark Twain: "Never invest in anything that eats." And, I might add, "Go south with the butterflies." During the cold grip of winter is not a good time to put your "finger in the dyke," as you should already be prepared.

Temperature, *humidity*, and *wind chill* are familiar terms. The following important terms are not as familiar:
The Thermal Neutral Zone (TNZ) is the temperature range (0 to 30 degrees C) in which a goat is most productive, with the least stress.

- The *Upper Critical Temperature* is the upper limit of the TNZ, above which a goat will become heat stressed.
- The *Lower Critical Temperature* is the lower limit of the TNZ, below which a goat will become cold stressed. This temperature is variable between individuals and is dependent on body fat, thickness of the hair coat and under coat, and diet. Neonates and the sick are unable to adapt.

The dietary energy intake must be increased during very cold weather. The Total Digestible Nutrients (TDN) must be increased by one percent for every degree centigrade below –15 degrees C. For example, a pregnant doe's TDN requirement is 55 percent, but when the temperature reaches –30 degrees C, her requirements increase 15 percent.

To avoid sudden changes in the energy portion of the diet (primarily grain), I recommend that you formulate the winter diet based on an average temperature of –20 degrees C, or 60 percent TDN. Sudden changes in her grain intake can lead to a mild to severe grain overload (rumen acidosis). The excess grain and acidosis can cause her to go off feed, which can lead to poor results. If she is in poor body condition, she may go down in the cold, primarily because she cannot maintain her body temperature.

The inability of the rumen to manufacture enough B vitamins, particularly thiamine, can cause polioencephalomalacia (sudden convulsions).

If her diet is low in calcium, a sudden stress can cause a calcium deficiency. This will cause her to become depressed and go down. To avoid this, make sure minerals and salt are provided in loose form and the mineral and salt feeders are full and in an accessible, draft-free location. If grass hay or straw plus grain are fed, a mineral with equal parts calcium and phosphorous should be available. If alfalfa is fed, a mineral with two parts Ca to one part P can be added.

If she is close to kidding, she may develop pregnancy toxemia, especially if pregnant with more than one kid.

You can take the following actions to prevent an acidosis from occurring.

- Have feed analysis done in the fall, and plan your winter feeding program well in advance.
- Increase the grain portion slowly so that the rumen can adapt.
- Feed whole grain rather than chopped grain.
- Have sodium bicarbonate available, either free choice in a feeder or mixed in the diet at 1.5 percent to 2 percent of the grain portion.
- Distribute the grain into several feedings per day so that the goats are eating no more than .25 to .3 kg per head at one feeding.

- If possible, feed your hay prior to the grain or put the grain on top of the hay if feeding hay in the bunk. If you feed the hay in self-feeders, make sure that they are never empty.
- Make sure there is adequate bunk space for all the goats to eat at one time. Boss goats, especially if they have horns, can be a problem.
- Fresh, clean water should be available at all times. The water bowls should be accessible, free from ice, and sheltered from the wind. If a group of goats suddenly stops eating or decreases their feed intake, always check the water bowls first, as adequate water intake is necessary for optimal feed intake. Do not rely on snow.
- Do not tub grind poor quality forage to reduce waste, as this does not increase the nutritional value of the forage. Because of the cold, the goats will eat the ground forage, but all it will do is decrease the nutritional value of the total diet.
- Clean the feed bunk daily and check for uneaten grain or pellets. If a lot are present, determine what they are. You may be putting feed into the bunk that your goats may not be eating it, instead only selecting what they like.
- As veterinarians, we often make the mistake of asking the client, "What are you feeding your goats?" and not "What are your goats actually eating?"

If your goats are not fed to meet their energy requirements during winter's cold grip, they will lose weight, as they will metabolically use their fat and muscle tissues to keep warm. This can lead to pregnancy toxemia, malnutrition, stress abortions, difficult births, and inadequate colostrum and milk production, which leads to starvation in the kids.

A Basic Guide to Ration Formulation

Do you need a fancy computer and a ration formulation program to create a ration for your goats? No, all you need are a cheap calculator, a pencil, an eraser, some paper, your feed analysis, and the basic nutrient requirements. Often computer-based ration formulation programs make a simple solution difficult by producing a lot of information that can be confusing to the novice. If you are proficient at managing spread sheets, you can set up your own ration formulation program.

You might be wondering what the point of a feed analysis is if you don't understand it. Take that feed analysis out of the recycling bin, and I will show you how to use it.

Step 1: Define your target production group

For the purposes of this example, let us select lactating does. This is applicable to meat does as well as lactation is important.

Average body weight: 50 kg to 55 kg
Target daily gain: 0 (if they are thin you may want to add a value for gain. This factor is mainly used for growing kids.)
Average milk production: 2.0 liters/day
Average fat percentage: 3.5 percent

Step 2: List the feeds you have on hand

Barley, second-cut alfalfa, green feed (oats)

Step 3: Make a table summary of the feed analysis

Use book or actual values. Most laboratories report two values for the feed: one as fed and the other at 90 percent or 100 percent dry matter. To compare silage to hay on an as fed basis is impossible, as the nutrients in the silage will be diluted by the water. Thus both feeds must have the water taken out or have an equal amount of water in them. For this example, all three feeds contain 90 percent DM.

Feeds	Moisture* percent	Dry matter percent	Crude Protein percent	TDN percent	Ca per-cent	P per-cent
Barley	0	100	11.2	73	0.07	0.29
Alfalfa	0	100	15	53	1.21	0.18
Green Feed	0	100	10	50	0.38	0.12

Step 4: Make a summary of the milking doe's requirements

	C. Protein (gm)	TDN (gm)	Ca (gm)	P (gm)	Dry Matter* Intake (gm)
Maintenance	73	530	2.1	1.4	2,000 (2.0 kg)

Lactation	136	684	4	2.8	250
Allowance**	0	0	0	0	0
Total	209	1,214	6.1	4.2	2,250

* This is calculated as a percent of her body weight. In this example, 4.0 percent was used as the average. For peak lactation she may eat up to 7 percent of her body weight in DM.
** Allowances are added to the formulation if the doe is thin or environmentally stressed. For example, in the winter the TDN and protein are increased relative to the temperature that the doe is exposed to (increase 1 percent for each degree below -15 degrees C). For winter feeding, add, on average, 5 percent to the protein and energy requirements.

Simple so far, I hope?

Step 5 & 6: The forage DMI should not be less than 40 percent of the total intake

For this example let us make the ration contain **60 percent forage**.
The forage DMI is .6 x 2.25= 1.35 Kg
The barley DMI is 2 - 1.2= 0.8 Kg or 0.9 Kg as fed.
Note: Because of the quality you should limit the green feed DMI intake to 1.5 percent of the doe's body weight, or .75 kg. Therefore .6 kg of alfalfa is needed.

Step 7: Calculate whether this simple ration meets the doe's needs

	DMI Kg/d	Amount as fed Kg/d	CP (gm)*	TDN (gm)	Ca (gm)	P (gm)
Alfalfa	0.6	0.7	90	438	7.3	1.1
Green feed	0.75	.83	75	375	2.9	.9
Barley	0.9	1	100.8	657	0.6	2.6
Total	2.25	2.53	266	1470	10.7	5
Require	2.25	2.53	209	1214	6.14	4.2
Balance	0		57	256	4.64	.39

* To calculate CP: alfalfa 0.6 x .15 = 90 g; green feed 0.75 x .10 = 75 g; barley 0.9 x .112=101 g

Congratulations! You have just evaluated a ration—or are you lost?

Step 8: Conclusions

The final ration will be as fed: alfalfa .7 kg; green feed .83 kg; and barley 1 kg.
All major nutrient requirements are met.
Mineral supplement should only contain phosphorous

Step 9: Variations

Feed only alfalfa hay. Use the green feed for the other groups.
If you decide to save the alfalfa and feed green feed because of quality, limit the DMI to 1.5 percent of the doe's body weight.

Add a protein supplement, e.g., 32 percent CP dairy ration (68 g), 46 percent CP soy meal (48g); 36 percent CP canola meal (61g). If you need the extra protein, feed the least expensive source grown in your area. These protein sources will also contribute to the nutrient balance. **Note**: Goats can sort out unpalatable feed, so always check the feed bunk for sorted feeds. For example, if you choose to top-dress with soy meal as the protein source, the deficiency in protein is 22 g, and soy meal contains 46 g CP in 100 g soy meal. One gram of soy meal contains 46/100 = 0.46 g protein; therefore, to meet the deficiency, you must feed 22/0.46 = 48 g. Based on the nutrient composition, 1g of soy meal will add .74 g TDN; 48 g will add 35 g TDN. Therefore, the mineral supplement only needs to contain phosphorous.

Once the ration is balanced for the major requirements, the main minor nutrients to check are selenium, copper, zinc, manganese, vitamin A, and vitamin E.

For example, if the mineral contains 16 percent P (16 g P in 100 g mineral), to make up the deficiency, calculate that [(100/16) x 2.1] = 13 g of required minerals.

At the end of this section, I have included blank ration evaluation tables to help you to calculate your rations. Ration formulation is not an exact science but is based on assumptions and choices. There are always several answers to the same question.

Most of the errors made in ration formulation are related to dry matter intake and moisture. The following example may clear the air.

How is a Canoe Trip Related to Goat Nutrition?

You are planning a two-week canoe trip, and you want to take fresh vegetables. You cannot take them fresh from the garden because you would need another canoe to carry them, so you decide to dehydrate them. This will reduce their weight considerably. You require ten carrots per day to meet your requirements. You estimate that ten carrots weigh 2.5 kg. (5.5 lbs)

After they are dehydrated and 88 percent of the water is removed, they weight 0.3 kg (equivalent to 12 percent DM). After your first attempt at eating the dehydrated carrots, you add the water back to make them easier to eat.

This is how rations are formulated. Your first step is to take all the moisture out when evaluating or formulating. You add the moisture back when you calculate the amount of feedstuffs "as fed" necessary to meet the group's requirements.

On the canoe trip, 0.30 kg (0.66 lbs) of dried carrots meets the ten-carrot requirement.

Potential Toxicities of Dietary Origin

Under most conditions animals will avoid toxic plants, unless they are eaten while grazing or fed accidentally, or unless the animals are starving. The following are examples of dietary toxicities.

Grain Overload

This occurs when your goat opens the grain bin (horns are a wonderful invention) and overeats. In most cases she will do this with her friend(s) to see who can eat the most, the fastest. If they gorge on whole grain, especially oats, the symptoms are milder than if the grain is chopped or rolled.

In order of severity:
wheat > hulless barley > barley > oats
chopped > rolled > whole

You may also be responsible for her overeating if you suddenly change the diet to more grain than she can handle.

Note: If they eat excessive protein, the clinical signs are roughly the same, but alkali instead of acid builds up in the rumen to complicate the picture.

a. What happens?

When an animal overeats on grain, acid starts to build up in the rumen. This kills off some of the microorganisms that cannot live in an acid environment. Thus acids start to build up in the blood. This results in the clinical signs.

b. Clinical signs?

These will vary depending on the amount and nature of the grain. Usually signs appear twelve hours after overeating. She is depressed, may nibble at hay, stagger, and might go down. She might be mildly to moderately bloated (swollen in the left flank) and dehydrated. She may scour or have loose manure with lots of grain in it.

c. Treatment?

Aim to neutralize the acid in the rumen before too much builds up in the body.

- Limit water intake to her requirements by hand watering. Do not restrict water.
- Give sodium bicarbonate (baking soda) diluted in warm water orally at 1 g/kg BW for three days. This is an excessive intake of bicarbonate but is well tolerated. One report states that the goats will recover within four days.
- If severe, you may a have to open her rumen (surgery) and remove the grain.

d. Complications?

- If pregnant, she may abort.
- The rumen may be permanently scarred when acid destroys the lining. This can lead to poor production and liver abscesses.
- Her feet may be sore, and she may appear to be walking on eggs (founder or laminitis).

Mycotoxins

These toxins are produced by fungi in feeds, primarily grain, but occasionally they will grow within a plant (especially fescue). Aflatoxin is the most common toxin, found in grain stored in a warm moist environment that supports fungal growth. Ergot is another mycotoxin, which appears as black grains in the heads of wheat, rye, and most grains under appropriate conditions.

Clinical Signs

Look for weight loss, decreased production and immune response (apparent as increased susceptibility to disease), abortion, and death. Ergot causes the tips of the ears and the end of the tail to blacken and fall off. The feet can also be involved in severe cases.

Plant Toxicities

Some of the plants toxic to goats are oleander, azalea, castor bean, buttercup, rhododendron, philodendron, English ivy, choke cherry, black cherry, laurel, daffodil, and lilies.

Super Phosphate Fertilizer Toxicity

Goats are curious animals, so any chemicals left lying around in bags or pails are quickly investigated by tasting them. Goats that have not had salt will satisfy their craving by ingesting fertilizer.

Clinical signs

Look for teeth grinding, diarrhea, depression, apparent blindness, and stiff-leggedness.

Excess Nitrates

Nitrites are found in shallow water supplies and drought- or frost-stressed cereal forages. Frost will damage the enzyme system that converts nitrates to amino acids. Until the enzymes repair, nitrites will build up in the plant. It takes between eleven and fifteen days for the enzymes to repair.

Errors in Ration Formulation

Errors can be made in the formulation or manufacturing of a commercial feed or supplement. If you suspect that this has occurred, stop feeding the diet immediately. If any animals have died or are ill, call your veterinarian to examine the animals immediately. The next step is to call the regional office of Agriculture Canada, and a field officer will come out to take appropriate samples for analysis. Most feed companies are cooperative, so notify their representative, who can examine the flow charts for that ration for any mistakes. Your final actions will depend on the number of animals lost, level of lost production, and the costs involved in investigating the case.

Urea Toxicity

Urea is fed as a source of nitrogen for the rumen organisms. They then convert the urea to proteins, which the goat digests and then utilizes for production purposes. Urea can be found in Dairy Concentrates, molasses/urea/mineral lick tanks, or fertilizers. Toxicity occurs when 30 to 50 g of urea/100 kg BW are consumed by a goat that has not been fed urea before. In goats, urea toxicity is seen most often when they are with cattle and a lick tank is put in for the cattle (the manufacturer often tells the client that it is safe for goats). Goats like molasses and thus may take in too much.

a. Clinical Signs

Signs occur rapidly after consumption, and death can occur within one hour. Signs include muscle and skin tremors, salivation, abdominal pain, bloat, convulsions, and death.

b. Treatment

0.5 to 1.0 liters of vinegar (household) via stomach tube or drench. Recovery can be quick.

The following are three cases (in sheep) to illustrate how errors can occur:

Case 1: Feed Tag Label Omission

Lambs developed white muscle disease despite being fed a commercially prepared grain ration with adequate added selenium and vitamin E (according to the feed tag). The feed manufacturer cautioned that no other sources of selenium should be fed. The cost to the owners of investigating this case was five hundred dollars. The final conclusion was that the feed company had old labels lying around which they wanted to use up. These indicated that selenium had been added, when in fact none had been added to that commercial ration. The feed company reimbursed the producer for the dead lambs and for the cost of the investigation.

Case 2: Excess Monensin Added to a Lamb Creep Ration

A commercial and a purebred producer split a ton of a commercially prepared grower ration. The commercial producer noticed that the lambs did not eat the ration and concluded that they would when they got used to it. The purebred producer wanted maximum gain on her lambs, so when the lambs refused to eat the ration she forced them by removing all other feeds for a few days. In about four days, one lamb died, and a number were stiff and reluctant to move. The muscles in their hindquarters were hard and painful. The initial diagnosis was white muscle disease. Blood samples indicated a normal selenium and vitamin E status. The only other cause for lesions of this type is Monensin toxicity. This was confirmed by Agriculture Canada. A new employee had just started on the feed mixing floor and accidentally added excess Monensin when he programmed the proportioner that added Monensin. The feed company covered all costs.

Case 3: Not Salt This Time?

This case involved the commercial producer from Case 2. After Monensin toxicity was confirmed, the feed company picked up the contaminated feed and delivered a new ration. The lambs refused this ration as well. The nutritionist was immediately called, and a sample of the feed was brought in. This feed contained about 10 percent salt. The flow charts indicated that normal levels had been added. The investigation revealed that this ration was produced using a new batch of calcium carbonate delivered that day. When the bin was sampled there was a lot of salt on the top of the calcium carbonate. The tanker truck that had delivered the calcium had not been cleaned out after the previous salt delivery. The result was contamination of the calcium. The only cost incurred was the collection of the contaminated ration and delivery of a (finally) palatable ration.

Ration Evaluation and Formulation: Do It Yourself Format

These blank tables are for you to practice ration evaluation and formulation.

Step 1: Define the production groups

Group	Age	Average BW	Target gain/day	Ave. Daily Milk Production Kg/day			Stressors
				Low	Average	High	

Step 2: List of feeds and supplements on hand and their costs

Forages	Grains	Supplements	Cost

Step 3: Summary of feed analysis

Feed	Moisture Percent	Dry Matter Percent	CP Per-cent	TDN Percent	Ca Per-cent	P Percent
	0	100				
	0	100				
	0	100				
	0	100				

Step 4: Summary of the daily requirements

	CP (g)	TDN (g)	Ca (g)	P (g)	DMI (kg)
Maintenance					
Growth					
Pregnant Yearling					
Pregnant/Dry					
Lactation (low, average, or high)					
Allowances					

Step 5: Determine the amount of forage to be fed

Note: Don't go below 40 percent of DMI. The DMI can vary depending on the environment conditions and the production group; it ranges between 4 percent and 7 percent of body weight.

Step 6: Determine the amount of grain and supplement to be fed.
(Isn't this easy?)

Total DMI - Forage DMI =

Step 7: Make a ration evaluation table.
* These are usually calculated based on the dry matter values, as the requirements are usually recorded as dry matter

Ingredient	Amount (kg)		CP (g)*	TDN (g)*	Ca (g)*	P (g)*
	Dry Matter	As Fed				
Total						
Requirements						
Balance						

Step 8: Identify problem areas and correct as economically as possible.
If you are totally frustrated, call a nutritionist for help.

The lab usually sends out a report with both the "as received" (fed) and "100 percent Dry Matter." Use the 100 percent Dry Matter to calculate nutrient balance and then convert the amount of dry matter required to the actual amount of feed as fed. Based on the protein and TDN, this is a poor to medium quality forage.

Table 13: Example of a Feed Testing Analysis Report of a Broom/Alfalfa Hay Sample

		As Re-ceived	100 percent Dry Matter Basis
Moisture at 135C	percent	12.5	0
Protein	percent	10.8	12.3
Sodium (Na)-Total	percent	.02	.02
Phosphorous (P)-Total	percent	.08	.09
Potassium (K)-Total	percent	1.36	.155
Calcium (Ca)	percent	.77	.86
Magnesium (Mg)	percent	.28	.32
Sulfur (S)-Total	percent	.116	.133
Acid Detergent Fiber (ADF)	percent	37.2	42.5
Nitrate	percent	Nil	Nil
Total Digestible Nutrients (TDN)	percent	39.33	44.94
Digestible Energy (DE)	Mcal/kg	1.73	1.98

Chapter 3

Keeping Your Goats Healthy

Necessary Supplies

On a regulate basis, do a reality check for all products you have on hand:

- check the expiration date
- make sure products are being stored properly
- ensure that the tops or seals are not broken or perished
- check that the color or the content of the product has not changed.

** **Note** Before purchasing any pharmaceutical products check the expiration date and only purchase those with the furthest away date.

The Medicine Cabinet

- An antibiotic: procaine penicillin will cover most of the common diseases you may encounter
- An oral electrolyte powder for orally treating mild dehydration related to scours
- Bottle of 50 percent dextrose for treatment of ketosis and pregnancy toxemia in does and cold, hypoglycemic kids. The response is best if given intravenously, but dextrose can be given orally or into the abdominal cavity (interperitional). For kids the IV dose is 12 cc of 50 percent dextrose diluted with 12 cc of sterile saline.
- A bottle of sterile saline
- 100cc of thiamine hydrochloride for the treatment of polio. This thiamine deficiency can occur if the goat(s) have been off feed for a few days, as associated with a change in feed, a sudden stress such as a weather change, lack of water, weaning, or moving to a new location. The goat appears blind, quickly followed by convulsions.
- A bloat treatment: For adults a product containing dioctol is best. For abomasal bloat or milk replacer bloat in kids, powdered household ginger, 1 tsp in 1/4 cup of warm water, orally is recommended.
- Betadine or an antiseptic/disinfectant wash such as Dettol
- Latex gloves and long plastic rectal sleeves to protect you from contamination by birth fluids, which can contain microorganisms that can affect your health. The fingers can be cut off and a latex glove can be put over the hand to increase protection and allow you to use your hands more effectively. The top of the rectal sleeve can be secured to your coverall sleeves with duct tape. Protect both your arms and hands.
- A lubricant
- 60 ml dose syringe or funnel
- Disposable syringes (3 cc, 5 cc and 12 cc) and needles (18-gauge, 1 and 1.5 inch)
- A small stomach tube for kids and a larger tube for adults
- A digital thermometer and stethoscope

Note: You may notice that I did not include a navel dip, as I did not use them when I had my own sheep. If the environment is dry and clean and the kids receive colostrum within few hours of birth, navel infections are unlikely. The navel should be dry within two days; if it is still wet and painful if pinched close to the body, antibiotic treatment is indicated.

Handling your Goat

To do a body condition score on your goats, or for any of the following management procedures, you will have to confine the goat. The following methods do not apply to free-roaming goats, including those in your neighbor's vegetable patch.

Goats must be handled on a daily basis, especially if you plan to milk them. From a young age, your goats should be conditioned to a regular daily routine. When feeding or milking your goats, bring them to a small holding area daily. They can be taught to lead using a small halter or a dog collar. If grazing space is limited and you do not want to put up a fence, then tether the goats to old tires, which are hard to drag quickly, to allow rotational grazing pattern. **Note: Make sure that they have water at all times, especially if being milked.**

Milking From Behind

It is the old-fashioned way—feed from the pail in the front, milk into the pail at the back. This works best for low and average producers. No danger of you or the bucket being kicked. But there are other perils, so your reaction time had better be quick.

Truth and Tales from Good Old Uncle Roy

Goats do not drive easily, as they tend to face their attacker. If pushed too hard or too closely, they may bolt and scatter in all directions. A snow fence or other portable fencing can be used to channel them. Dogs must be well trained to stay fifty to a hundred feet behind.

Body Condition Score (BCS)

Body condition scoring is a valuable management skill for most livestock producers. BCS is a visual and hands-on appraisal of the muscle mass and amount of subcutaneous fat. Dairy goats are difficult to score, as they store more fat in the abdominal cavity than under the skin. Because of this, BCS on goats is done by examining the brisket as well and averaging this value with the traditional lumbar score.

The best times to BCS goats are when:

- drying off, when an ideal score ranges from 2 to 3.5
- kidding: ideal score 3 to 3.5
- forty-five days into lactation: ideal score 2.0. She should not drop more than 1.5 points during this period.

Traditional Lumbar Body Condition Score

This is a hands-on evaluation of your goat's body condition. The site evaluated is the loin area just behind the last rib and over the ribs on both sides. Feel along the back for the tip of the back bone or spine (spinous process) and then the space between the top of the spine and the part of the back bone (transverse processes) just above the flank area. Judge the fullness of the muscle tissue under the skin and the amount and softness of the fat just under the skin. Then run your hands over the rib cage to evaluate the fat cover under the skin.

Brisket Score

This is a hands-on evaluation of the brisket the area between the front legs. Run your hands down over the rib cage and then between her front legs. Evaluate the fat cover and the amount and firmness of muscle tissue under the skin between her front legs.

Table 14 Body Condition Score

Body Condition Score	Lumbar Site	Brisket Site	Significance
0	Hide stretched over the skeleton	Emaciated, bones of brisket and ribs easily felt Large, firm callus present	Individual animal: chronic disease; bad teeth Group problem: starvation
1	Still very thin, some evidence of muscle tissue	Still skeletal Articu-lations rounder Callus present but movable	
2	Thin, some muscle and fat present Skeleton is obvious	Articulations dif-ficult to feel Fat pad under the skin and between muscle Callus absent or small	Look at nutrition
3	Firm muscle, 1/4-inch of subcuta-neous fat Ideal for breeding	Distinct ridge be-tween mass of fat, muscle, and the bone	Ideal
4	More fat evident under the skin and over the ribs	Fullness in brisket area	Over-feeding
5	Obese	Fat tissue obscures the underlying muscle and bone	Concern

Caution: If you are a new producer or have any concerns about doing this procedure, call your veterinarian and do a supervised hands-on practice of the technique together.

Navel Dipping

At birth, dip the navel in 2 percent iodine, teat dip, or iodine-based disinfectant. This has been traditional for years, yet I now see on some farms a high incidence of navel ill in spite of navel dipping. I stopped dipping our lambs' navels as an experiment and found the only lambs that developed navel ill (umbilical infection) were those that got off to a bad start. Some does may kid in the spring muck so the kid's navel becomes covered in dirt and manure, but if she is a good mother and encourages the kids to get up and nurse, the navel will dry, and no problems are likely to occur. Navel dipping does not appear to help neglected lambs. The best solution is to cull the mother. Most bacteria in the environment are not harmful, and the doe is likely to be immune to any harmful microorganisms present in her environment. The most important point is to be sure the navel cord is completely dry within two to three days after birth.

Identification

The kids should be identified either by physical description or an ear tag. Purebreds are tattooed in the ear with farm letters, a number, and the year letter. A new method is to place a microchip under the skin at a specific location. A mandatory Canadian National Identification ear tag system is now in place so that the origin of goat can be determined, no matter how many owners it has had.[7]

Castration

All males not destined to be herd sires should be castrated. A castrated buck is called a wether. If they are not castrated, separate them from the females at twelve weeks of age.

The ideal age for castration is before one week, no later than one month; the earlier the better. You must check to make sure both testicles are present in the scrotum. If the kid is older than two weeks he will require an anesthetic (goats are very sensitive to pain).

You must ensure the kid is protected against tetanus, either through its vaccinated mother's colostrum or by an injection of antitoxin. **Note: This is important in areas where tetanus is a problem.**

Restraint for Castration

The kid is set so its rump is placed on a solid base. The hind legs are brought towards it head and held along with the front legs

Burdizzo**

**Use this method after instructions from an experienced person
A small burdizzo is expensive and difficult to use until you find the structures and know how to manipulate the burdizzo. Isolate the cord (about the size of a round boot lace) just above the testicles. Hold the cord in place, close to the outside skin. Place the open burdizzo across two-thirds of the scrotum just below the rudimentary teats. Check that the cord is included and then close and crush the cord. Hold for ten to fifteen seconds. Repeat on the other side. This cuts through the skin and separates the blood supply and spermatic cord from the testicles. The cords must be cut completely through, but do not cut across both cords at once.

It is advisable to crush the cord twice. Swelling may occur in the next twenty-four hours but should disappear over the next two to three days.
Caution: Clients have crushed the small penis instead of the cords. Make sure the structure you are crushing is attached to the testicle.

Open Castration

In this method, the testicles are removed through an incision in the scrotum, which must be large enough to allow drainage.

Grasp the lower end of the scrotum. Push the testicles up towards the body, out of the way of the scrotum. With a sharp knife cut the lower one-third of the scrotum off.

Identify and grasp the testicle and pull it slowly downward until the testicle and the cord are outside. Gently pull until the cord breaks. The portion of the cord not broken will go back in. No cord should be evident outside the scrotum.

Put kids in a clean, dry, well-bedded pen. Watch for swelling.

Elastrators

You will need a special applicator and the proper size and type of rubber band—a rubber band from your office is not appropriate. (Tetanus has been a problem with this method.)

Place the elastic band securely above the testicles, cutting off the circulation to the testicles. Be sure that both testicles are included. If a testicle is left in the body the goat will act like an intact male at maturity. Watch for swelling and pain. The scrotum and elastic should fall off in a week, leaving healthy tissue underneath.

Some producers will push the testicles up into the inguinal ring and then place the elastrator around the base of the scrotum. These animals may grow well but at maturity they will develop the traits of a male.

Disbudding/ Dehorning

Disbudding is the removal of the horn buds on a young goat; dehorning is the removal of grown or growing horns on an older goat. This procedure is not without risk of death or infection and must be done by a veterinarian. The practice of either disbudding or dehorning is controversial. Some people believe that goats should be left in their natural state, while others believe that horns can cause problems. Range goats or goats used for fiber production are not usually dehorned. In either case, dehorning is a serious procedure that requires experience to perform. The horns act as a cooling mechanism and connect directly to the sinus cavities in the skull. If you are new to goats or are contemplating getting goats, consider the subject of horns very carefully before making your decision on whether to get horned goats or polled (hornless) goats. I advise new producers to disbud as goats with horns can be bullies.

Balance and Symmetry Are All-Important

"Mrs. Ross was a ruthless tyrant, until a day in battle, when the end of one horn broke off,

whereupon she was beaten, hounded, and tormented mercilessly by other does. She looked awful, and a lop-sided horn did not help her appearance, so I trimmed the horns to equal length. She was restored to symmetry and balance. She won the next election and returned in triumph to her place as a ruthless tyrant".

Truth and Tales from Good Old Uncle Roy

Billy Goats Are Demolition Specialists

"They enjoy rearranging the scenery. Old Father Casey sometimes found it boring to be tethered on that long chain near the steel Quonset, so he wiled away the hours bashing the corrugated steel panels of the walls. The bashing made a glorious boom. Neighbors thought it was artillery practice at the army base. The walls yielded bit by bit to the onslaught. The Quonset looks now as if a bulldozer went wild on a Saturday night".

Truth and Tales from Good Old Uncle Roy

Horns or No Horns

"At one stage the fabulous Blucher Greys included both horned and polled animals. And that was a lesson learned the hard way. A bald pate is no match for horned glory. The polled animals were bottom of the heap. One was X-rayed by a well meaning vet for a problem, real or imagined. The plates revealed that every rib had been broken (and painfully healed). I was accused of beating her with a two-by-four. I would never do such a thing! It was the price she paid for being omega goat in a herd with horned boss ladies. Keep them all horned, or all polled, but not both. You cannot have your cake and eat it too".

Truth and Tales from Good Old Uncle Roy

Goats should be disbudded within the first two weeks of birth (preferably between two and four days). To dehorn adults requires an anaesthetic as they are sensitive to pain. In goats there is very little space between the base of the horn and the brain. Dehorning the "boss" goat can result in the other goats beating her up, and I have seen some cases where she has been killed.

Disbudding Iron

If using a calf dehorner, check to see if it will cover the horn bud. The disbudding iron should be a minimum of 200 watts with a 3/4- to 1-inch-diameter tip. In one study, barring equipment malfunction and operator fatigue, fifteen to twenty kids from two days to five weeks of age can be disbudded in one hour.[8] The disbudding iron should be used once horn buds can be identified.

Some producers restrain the kid in a special box that leaves only the head exposed.

Horned kids have two tufts of hair where the horns will be. Under each tuft is a bald spot. Clip the hair to expose the horn buds. Heat the iron until cherry red.

Infiltrate under the skin around the base of the horn with 2 percent Xylocaine.
Note: 2 mg/kg BW of Xylocaine is toxic to kids, so consult your veterinarian about the proper dilution, as you must be careful.

In young bucks, burn the skin just behind the horn bud toward the center of the head (an area of thick dark skin). Apply just enough pressure surrounding the base of the horn to hold the iron on the skin. Leave the iron in place for four to six seconds. The skin around the base of the horn bud should be copper-colored. The horn bud should be removed easily and quickly. Do the next horn bud. Avoid blistering the skin as this may cause brain damage.

Other Methods of:

Disbudding

- Elastic bands can be used to remove the horn buds, but if improperly applied, they can result in scurs (deformed new horn growth).
- Caustic paste is used to remove horn buds. To apply this paste properly, clip the hair around the horn bud, put a round bandage over the bud, and smear Vaseline around the edge of the bandage. Remove the bandage and apply the chemical. The Vaseline will act as a barrier to prevent the paste from running into eyes. Restrain the kid for thirty minutes until the burning sensation stops, so that she doesn't rub her head. **Caution**: The paste can run into the eyes and burn them; if improperly applied, it can leave scurs and burn the skin.\

Dehorning

- Sawing or gouging horns requires an either a local or general anesthetic.(which must be administered by a veterinarian)Goats have two separate nerves to the horns (cattle have only one) both nerves must be frozen to avoid pain during dehorning

Routine Procedures

Passing a Stomach Tube[9]

A stomach tube is used to deliver oral medication and fluids and to relieve bloat. Being able to pass a stomach tube may one day save a life.

Neonates

Stomach tubes are used to feed and/or give electrolytes to kids that are not eating. The stomach tube should be flexible, less than 1.5 cm in diameter, and made of soft plastic or rubber. The end going into the stomach should be rounded, with holes at the end and on the side of the tip (check with your veterinarian or local hospital or plumbing supply stores).

1. Lubricate the tube with water or a lubricating gel.
2. Place the kid on its right side and measure the amount of tube that is required to pass from the mouth to the last rib. Mark the tube at this point.
3. Hold the kid on your lap in an upright position. Gently open its mouth and pass the tube over the tongue and into the esophagus.
4. Watch on the left lower side of the neck; you can see the tube passing down the esophagus, and you can also feel it. Insert the tube until the mark you made is in the mouth.
5. Blow into the tube and feel or hear the air bubbles going into the stomach on the left side behind the ribs.
6. Attach the liquids to the end of the tube using a 60 ml (cc) syringe or a funnel and allow it to flow by gravity into the stomach. Forcing the fluids may rupture the stomach or cause fluid to back up into the lungs.

7. Feed small amounts—50 ml (cc)/kg body weight at one time, three to five times day. Digestion of milk slows down when the kids are hypothermic.[10]
8. After all the medication is delivered, remove the funnel, pinch the top of the tube, and withdraw it slowly.
9. Clean the tube and funnel thoroughly between kids.

Adults

The technique is almost the same as for the neonate, but a larger, 2 cm diameter tube is used. The procedure is rarely required but is used to relieve free gas bloat or when oral medications are required. A piece of PVC pipe about six inches long with a diameter large enough to pass the stomach tube through is required to prevent the goat from cutting the tube with her molars.

1. Restrain the goat's head and stand beside her.
2. Place the PVC pipe into her mouth and over the tongue.
3. Lubricate the tube with Vaseline or water. In the winter you will need to warm the tube.
4. Pass the tube into the pipe and gently into the esophagus. Often you can see the tube passing down the left side of the lower neck.
5. To check if the tube is in the rumen, blow into the tube and listen for bubble sounds in the left flank. Also check for the smell of rumen contents through the tube.

Drenching

Drenching can be used to give oral medication. The most common way is to use a narrow-necked soft drink bottle or a 60 cc dosing syringe. Drenching guns are used if a large group is medicated, but be cautious! Only use the gun designed for the size and age of the goats treated, and do not ram it into the back of the mouth. (I have seen this done).

1. Tilt the head up at a 45-degree angle. Place the opened end of the bottle in the cheek pouch on either side of the mouth.
2. Pour slowly, as they can aspirate the liquid into their lungs if not allowed time to swallow.
3. Liquids such as mineral oil that have no taste do not always stimulate a swallowing reflex and can go into the lungs with fatal results. To solve this problem, add a flavor to the oil (e.g., a teaspoon of powdered ginger).

Injections[11]

Always read the manufacturer's directions. Injections can be given just under the skin (subcutaneous), into the muscle (intramuscular), into the vein (intravenous), or into the abdomen cavity (intraperitional). Injections should be given where the hair and skin are clean, free from dirt and manure. The hair can be clipped away, but using alcohol or disinfectant to clean the site is not necessary, as these products may cause irritation and release bacteria on the surface of the skin.

Subcutaneous (under the skin)

This method is used for most vaccines. Pick an area relatively free of hair and with loose skin (the flank, over the ribs, or the neck). Never inject more than 10 cc (ml) at one site.

1. Lift the skin in a tent-like fashion.

2. Insert the needle at an angle almost parallel to the skin.
3. Pull back gently on the needle to make sure you are not in a blood vessel. If the syringe starts to fill with blood, withdraw and start again.
4. Smooth the material under the skin.

Note: Avoid injections under the elbow. Large nerves to the leg are in this area; if nerves are damaged, the animal will drag its leg. This can be permanent or temporary.

Intramuscular (into the muscle)

Any large muscle mass will do (if you can find one on a goat). The preferred area is in the lower part of the neck: the triangular area just in front of the point of the shoulder. Other areas are over the rump or the back of the hind leg. Avoid the hind leg (a prime cut) in market animals. Injections into the hind leg can cause sterile abscesses and scarring of the muscle tissue, which is unacceptable in meat animals. Do not inject more than 10 cc into one site. It is unlikely that you would ever require more than 10 cc of any IM preparations. Some of the products used can be quite irritating, and the goat may have severe discomfort after the injection.

1. Use an 18-gauge, 1- to 1 1/2-inch needle and insert into the muscle.
2. Pull back on the plunger to be sure you are not in a blood vessel.
3. If using the hind leg, direct the needle towards the tail to avoid the major nerve.

Note: Avoid injections under the elbow, large nerves to the leg are in this area; if damaged, it will cause the animal to drag its leg. This can be permanent or temporary.

Intravenous (into the vein)

This is an ideal route for many reasons, especially with products that require an early high blood concentration or if intravenous fluids are needed. **Caution: Only use products that are marked "for intravenous use."**

This procedure is difficult to describe and requires a live hands-on demonstration. The jugular vein is used as the injection site. The jugular vein runs along both sides of the neck along a groove just below the bottom of the bones of the cervical vertebrae and just under the skin.

1. Clip the hair in this area to make the jugular vein more visible.
2. Hold the vein off by putting pressure in the lower neck just in front of the shoulder. You should be able to see the jugular vein fill with blood and be able to feel it. (This is easier if the hair is clipped over the jugular.)
3. Insert the needle (18- to 20-gauge) gently into the vein at an angle. Blood will come out of the needle if it is in the vein.
4. Place the syringe on the needle and pull back to make sure you are still in the vein.
5. Slowly inject.

Note: If you are in an artery, the blood spurts out and is bright red. If this occurs remove the needle at once as medications should not .be injected into an artery. If you are unsure do not inject.

Intraperitional Injection

This site is used when you cannot do an intravenous injection. An 18-gauge needle is inserted into the abdominal cavity one inch above the navel, behind the sternum and directed towards the tail. Fluids can then be put through the needle. When fluids enter the abdomen, they are taken into the body quickly. Irritating or very concentrated solutions must not be used, as they can cause chemical peritonitis. Only products that can be used intravenously should be considered for Intraperitional injections. If bacteria are introduced into the abdomen, peritonitis can occur. I have not used the site because I prefer the intravenous route. For a lay person, if care is taken, this procedure can save lives.

Hoof Trimming[12]

Hoof trimming is quite easy and should be done on a regular basis. Regular small pruning shears, foot rot shears, or a sharp hoof knife can be used.

First remove excess horn growth from the wall.

Then trim the sole and heel to meet the wall. Care should be taken not to go too deep. Murphy's Law always applies: trimming the hooves just prior to an important event will always cause lameness.

Note: Check the feet closely when examining a lame goat. Examine between the claws (interdigital space) for heat, swelling, and redness. Scent glands can become infected and resemble foot rot.

Procedures for examining manure for parasites[13,14,15]

There are two procedures that can be used for examining manure for parasites. Before doing these procedures, consult with someone with experience in identifying the eggs of parasites.

Procedure 1: Looking for eggs
Procedure 2: Counting how many eggs are in a gram of feces

Table 15 Procedures for Examining Manure for Parasites

	Equipment	Feces	Floatation Fluid	Procedure	Recommendations
For both Procedures	Microscope (100X power) Fine-screen sieve or tea strainer, Plastic mixing cup Disposable coffee stir or popsicle stick Labeled Ziploc bag for feces	Fresh feces just passed or taken from the rectum, from individuals or mixed together from several goats in the group.	Dissolve 220 g sugar in 3/4 cup of boiling water. The specific gravity should be between 1.20 and 1.335 (a hygrometer can be purchased at any auto supply store)	Completely mix the feces with the solution. Once well mixed, pour through the sieve or tea strainer into another cup	
Procedure 1	A 35 mm film canister or equivalent small container Glass slides and cover slips	8 to 20 fecal pellets	Half fill mixing cup	Pour strained liquid feces into a film canister. Allow to overflow a bit. Place a glass cover slip on top and let stand for 20 minutes. Place the glass cover slip liquid side down onto a glass slide and place slide on a microscope	Used more often for screening a group of goats for parasites. Decisions on deworming will depend on the type and number of parasites eggs seen, the time of year, the condition of the goats, and their future movement.

Procedure 2	A small scale McMaster fecal egg counter	Two grams of feces	28 cc added to mixing cup	Fill an eye dropper with the well mixed solution and feces. Fill the first counting chamber and then the second chamber. The slide is then placed on a microscope and viewed at 100X. Find a corner of the counting chamber and begin counting eggs Calculate eggs per gram (EPG) # eggs in left chamber + # eggs in right chamber = total eggs x 50 = EPG	Deworm if: Buck and dry does >/= 2,000 EPG Lactating does >/=750 EPG All others >/= 1000 EPG

Can I do my own postmortems?

This section is not for the squeamish, the faint of heart, or novices. My answer to this question is a weak maybe. My reservations are numerous, but the main ones are:

1. You can spread disease to your other goats by exposing them to the blood and fluids of diseased goats.
2. Some diseases can be transmitted to you if care is not taken.

Because goats can rapidly decompose, a postmortem should be done within the first few hours after death. In many instances, this is not possible for you to do. Post-mortems are also expensive, and often you have to drive many miles to an appropriate laboratory. Any goat that leaves your farm dead or alive must be identified through the National Livestock Identification Programme.

Requirements Needed To Do a Postmortem

Knowledge and understanding of the normal internal anatomy are needed. This information can be found in books, or if you slaughter your own goats you can observe their anatomy. If your goats are slaughtered at the local abattoir, then you should go at the time of slaughter and become familiar with the normal anatomy of goats and what normal organs look like. A digital camera is a valuable tool for recording your findings so that they can be sent to a veterinarian for examination and interpretation. I would advise you not to send it to America's funniest home videos.

Choose a site for opening the animal far enough away from live animals so they will not be exposed to the blood and fluids. If the postmortem is done in a shed or on the back of your half-ton pickup, the area must be thoroughly washed and disinfected. Do not wash your truck box in a local stream or dugout. Disposal of the carcass can be a problem. Cremation might be the best route.

Always wear protective clothing, which includes a plastic apron, washable boots, protective eye glasses, a protective face mask, and rubber gloves. Disposable coveralls could also be worn.

A sharp butcher or a skinning knife is essential. A dull knife from your kitchen is inappropriate and dangerous. A meat saw is useful for cutting bones, but it should only be used for this purpose or be thoroughly cleaned and sterilized before it is used to cut up a healthy carcass.

Other equipment you should have on hand is scalpels, surgical scissors, and forceps.

Plastic bags of different sizes are useful. Leak proof (Ziploc) bags are ideal for sending tissues to a diagnostic laboratory.

A cellular phone can be useful; on several occasions I have instructed clients over the phone on how to proceed with a postmortem.

Steps to Follow

Often the history and clinical signs will help you decide if a postmortem is necessary. If a disease of the nervous system is suspected (example: scrapie or rabies), then the animal or its head must be sent to a diagnostic laboratory. **Note ** Contact your veterinarian—do not do your own postmortem. If you suspect that the goat has a disease of the central nervous system, do not shoot it in the head.**

Step One: Place the goat on its left side. Evaluate the body condition. Check any external swellings or injuries for evidence of scours or abnormal discharges. Examine the condition of hair and skin, and check the udder for abnormalities. If it is a buck, check the testicles for abnormalities and the hairs around the prepuce for small bladder stones. Check the end of the penis for stones and along the penis for any swellings. Examine the mouth, teeth, tongue, and eyes.

Step Two: Starting under the jaw, place the knife blade under the skin and slit open the skin down the neck, over the sternum and the abdomen, to the pelvis. Avoid cutting into the abdomen, which may be difficult if the animal is bloated or badly decomposed. If you accidentally cut into the rumen, the sudden release of pressure can cover you and many observers with the rumen fluid. The smell can linger for several weeks.

Step Three: To examine the chest cavity, strip back the skin and cut the muscles attaching the front leg to the chest. Lift the leg to expose the ribs. Are the tissues dry? Is there any fat under the skin? How much? Split the sternum with a meat saw to open the chest cavity. As you open the chest, observe if any fluid escapes; note the color and if it is clear, cloudy, or bloody. The lungs should be examined next. Normally they should be spongy and pink. After an animal dies, the lung becomes dark in spots but still feels spongy. These changes are most notable in the lung that was facing down when the animal died. Abnormal lungs can be dark and firm, and only portions of the lung may be involved—generally the front part of the lung. Adhesions may be evident when the chest cavity is opened. The surface of the lung may be covered with fibrin, and there may be abscesses within the lung itself. The next organ to examine is the heart. The heart is enclosed in a thin sac. Open this carefully. Normally a small amount of fluid is present. If there was a problem with the heart, the sac may be filled with fluid and the heart may be enlarged.

Step Four: Now remove the lungs and heart from the chest. This can be done by cutting the windpipe and associated tissues where they enter the chest. Strip the tissue along the backbone until the heart and lungs are free. Check the other lung and the outside of the heart. The heart is difficult for an inexperienced person to examine. Open up the windpipe to check for abnormal fluids, pus, or parasites. If you want to determine if what you are seeing is abnormal, place the tissues in well-sealed plastic bags, keep refrigerated, and send them to the laboratory as quickly as possible. Each bag should be labeled with your name and address, the goat's identification, age, and gender, along with the type of tissue.

***Step Five**:* Now examine the abdomen. Because of the large amount of bacteria present, the abdominal organs change color and decompose rapidly. Open the abdominal muscles and fold them back with the skin. As you do this, check for abnormal and/or excess fluid. The first organ encountered is the abomasum, or a true stomach. Beside and above this organ are the liver and the small gall bladder. A gall bladder larger than a golf ball suggests that the goat has been off feed for awhile. The liver should be brown to purple in color with sharp edges. If a doe died from pregnancy toxemia, the liver will be enlarged and yellow and brown in color, with rounded edges. Abscesses in the liver are an indication of *caseous lymphadenitis*. It will be difficult for an untrained eye to identify many of the abnormalities that may be present. The amount of abdominal fat present around the kidney and the intestines should be evaluated. Emaciated goats will have very little, if any, fat. If you detect anything unusual, place it in a labeled Ziploc bag, cool it, and send it to a diagnostic laboratory immediately.

Kid Necropsies Special Features

A difficult or stressful birth should be suspected if the kid's hair is stained yellow or orange. If the eyes are white or cloudy, the kid died in utero. Open the chest and remove a piece of lung and then drop it into a cup of water. If it sinks, the kid had not breathed; if it floats, the kid was born alive.

A kid that died from starvation will have no fat present in and around any tissue. The abomasum will contain no milk, and the urinary bladder will be large and full of urine. If the kid died from scours and/or dehydration, the bladder will be small and empty.

Note: Record your novel approaches to these techniques.

Chapter 4

Breeding Goats[16,17,18,19,20]

Introduction

Hormonal manipulation of the reproductive cycle can improve efficiency if properly used by a knowledgeable producer. Rapid advances in reproductive technologies are occurring. Some are controversial, particularly when applied to human reproduction. Goats are ideal animals to use because of their small size and diversified commercial value. Use of these technologies could lead to rapid genetic improvement. However, they are expensive, and selection for one strong trait may compromise another (e.g., trait for rapid growth can lead to abnormal bone development and chronic lameness). When manipulating the reproductive cycle, always check with a specialist in this area, as new drugs and methods may be available.

Puberty

Goats can reach puberty at a young age. The exact age depends on the breed. Most breeds reach puberty by five months of age. The Angora goat may not show sexual activity until its second year. Goats should not be breed until they reach 60 percent to 75 percent of the mature body weight for the breed's standard.

Breeding Season

Goats are seasonal breeders and respond to decreasing daylight ("short day breeders"). Both females and males respond to the photo period. The breeding season starts late August and continues until March. The peak breeding period is in October and November. Early in the estrous period, does may short cycle (in seven days). Approximately 55 percent of these cycles are anovulatory: no eggs are produced. If you raise goats at the equator, estrous occurs throughout the year.

The rut of the buck will affect the occurrence and detection of estrous in the doe. Thus the anestrous (non-breeding) period is often longer in small groups isolated from bucks or where the buck is de-scented. A warm cloth saturated with buck scent kept in a closed jar and then opened around does may help in estrous detection. A doe will stand for a buck only when in heat.

Estrous Cycle and Behavior

The estrous cycle lasts from nineteen to twenty-one days. Extremes can occur when the buck is first introduced to the group. The length of the estrous period (standing heat) is approximately thirty-six hours (twenty-two hours in Angora goats). This period is also shorter in young does. Ovulation occurs a few hours after the end of standing heat, so breeding should take place near the end of the estrous period. For maximum conception, does should be mated with the buck at the onset of standing heat and at twelve-hour intervals until it subsides.
If the producer is using artificial insemination or is hand-breeding, the does should be checked for standing heat several times a day. A doe in standing heat exhibits a typical behavior. If the producer places a hand and applies pressure on her rump, she will stand and flag her tail. The

owner must keep track of the onset and length of the estrous period when deciding to breed.

And More About Perfume

"I threw an elegant soiree. All the best people were there. And so were my ordinary friends (a salesman and a farmer). These two buddies decided that Brother Jean Baptiste should join the party. It was blossom time. I met the threesome at the kitchen door and evicted them. After all, admission was by invitation only. After I refreshed their drinks, they left. One was wearing a new suit, bought for the occasion, but also to look good on his first real job as a feed salesman. He told me later that he had his suit dry-cleaned three times, but still had to burn it. He learned that feed salesmen in blue jeans do okay".

Truth and Tales from Good Old Uncle Roy

The Perfume of a Billy in Full Bloom

"*Nothing on earth can match its power and glory, pungency, penetration, taste sensation, and violation of all senses of mankind. I frequently had to travel as part of my job, which required the services of a caregiver who could pinch-hit. (Milking goats is not a skill possessed by many.) She was resourceful, capable, supportive, loyal, and fazed by nothing—she was wonderful. I was due home on the late plane, so she finished her job, changed into grubbies, and drove out to do" chores. She did the chores, had a bath, and changed clothes. She drove to the airport. I stepped off the plane onto the ramp and knew she was there to welcome me home".*

Truth and Tales from Good Old Uncle Roy

Buck Behavior

Because of their unique smell and disgusting behavior during breeding season, the buck is often avoided except to feed. Bucks establish a social hierarchy, the alpha buck being dominant. This can be a problem if the alpha buck has low fertility. Fortunately most operations are small and only one buck is needed, unless AI is used.

Did you know?

The buck's delightful odor (pheromone) is linked to fleece lipid (fat) production. This lipid production is maximized by a good diet. Do not use this as an excuse to starve him.

Fertility in bucks is seasonal. They return to normal sexual behavior two weeks prior to the does. The goat odor serves as a stimulus for the does to start cycling. In response to the photoperiod, the buck's semen quality and quantity increases. This response varies among breeds.

Gestation Period And Litter Size

The average length of gestation (pregnancy) for the goat is 150 days. Placentation (attachment of the placenta to the uterine wall) is complete fifty-two days post-conception; the pregnancy is dependent on hormones produced by the corpus lutem (the yellow body on the ovary that forms after an egg is released). Ninety percent of deliveries should occur between 146 and 154 days. Larger litters tend to have a shorter gestation period. Litter size increases in the second and subsequent pregnancies. Nubian goats have the highest number of kids and

Angora goats the lowest.

These numbers represent averages and are not written in stone, and none of the goats have read them, so some variations may occur in your goats. These may be normal for your group, but if the does are ill or failing to conceive, consult your veterinarian

Kidding

(Some trivia you can impress the new producer with.)

In one study:

- 27 percent of the births occurred with the doe standing and 73 percent while she was recumbent.
- 81 percent of the births were a normal delivery, 15 percent were breech, 4 percent with forelegs first/head back.
- The average length of active labor was twenty minutes for singles and thirty minutes for twins.
- Between 33 and 50 percent of the does partially ate the placenta (which can rot in the rumen and cause indigestion). This is an instinct to remove the evidence of a recent birth.
- By two hours postpartum the doe accepts her kids; a strong bond develops within four hours, which does not seem to be related to nursing or licking.

The placenta should pass within a period of twenty minutes to five hours. The postpartum uterine secretions from the vulva vary from none to 200 ml of bloody discharge. These secretions should be significantly decreased by sixty hours and ceased by five days. By twenty-eight days postpartum, the reproductive tract should have returned to its initial nonpregnant state.

Stillborn kids account for 4.6 percent of all kids delivered. Entire litters can be stillborn. Inter sexes represent 1.5 percent of the kids born. Dystocia (difficult kidding) occurs in approximately 3 to 5 percent of births. This information gives the producer some guidelines to compare to. For example, if the incidence of stillbirths is 6 percent, veterinary assistance should be sought.

Goats appear more susceptible to abortion than other species. The main causes are infections and nutrition (see section on fetal wastage). Habitual aborters (usually older does) should be culled. A low level of stress abortions is common in Angora goats. This is related to the competitive nutritional demands of late pregnancy and fiber production. Fiber production usually wins.

False pregnancies can occur in goats. Does gain weight and appear pregnant because sterile fluid accumulates in the uterus. When this fluid is expelled around the expected day of kidding, old timers called it a "cloud burst." (You are right in saying she wasn't kidding.) When diagnosed, this condition can be terminated with prostaglandin or oxytocin.

Manipulation Of The Breeding Season

This section serves as a brief overview of the methods used to manipulate the reproductive cycle. Most breeders prefer to let nature take its course. These technologies when used by an inexperienced producer can be financially hazardous, if not dangerous, as all these hormones are used in human reproduction—"It's not nice to fool Mother Nature." A veterinarian who specializes in reproduction should be consulted.

Estrus Synchronization

- Pessaries (vaginal) of flurogesterone acetate (30 mg) or medroxyprogesterone acetate (60 mg)
- Remove in twelve to fourteen days
- Inject with 5 mg of PGF 2 alpha
- Estrus started twelve hours sooner in the flurogesterone group.

The Buck

Introduction of a rutting buck to isolated does will induce a reasonably synchronized estrous. Vasectomized bucks can be used in this role. Bucks have a profound influence on the estrous cycle when introduced to the does in late anestrous. In one study, fifteen of seventeen does showed behavioral estrous seven days (plus or minus one and a half days) after the introduction of the buck. Three of these does had short cycles followed by normal cycles.

We Are Pleased to Provide a Smoke-Free Environment

"Goats love tobacco, but they don't smoke. Some people should be so fortunate. But goats may suffer yet from the Canadian preoccupation with what is politically correct. Maybe one day perfume will be forbidden in public places and restaurants. What is poor billy to do? Here is how we handle it in the Land of the Blucher Grey. In those frequent years when demand for breeding stock and replacement females is less than brisk, we worry not about the daddy of the new crop, since he will be shish-ka-bob anyway. So we keep back a handsome billy kid in the spring. By fall he is fully capable but still odor-free. He loves them all—the long, the short, and the tall. Then he goes down the road to sire late arriving kids for the petting zoo. And then he goes to the auction for shish-ka-bob with a little curry or chisnik. I get the check, perfume-free".

Truth and Tales from Good Old Uncle Roy

Hormones

Prostaglandin (PGF)

Natural prostaglandin causes the breakdown of the corpus lutem (yellow body) that forms to fill up the hole in the ovary after ovulation. They signal to the goat that she is out of heat. When PGF is injected into a doe, standing estrous occurs within two to three days. Thus, commercially produced PGF can be used by producers to control the estrous cycle and to plan insemination. When given, PGF successfully induces the estrous period (standing heat) between days four and seventeen of the estrous cycle. After the use of PGF a small percentage of does may short cycle; i.e., return to the estrous period ten days after the induced estrous. These are then followed by a normal cycle.

To ensure that all does respond and estrous is synchronized, two injections of PGF can be given eleven to fourteen days apart. The does are then bred when the estrous period is observed, about fifty hours later. As little as 1.25 mg Lutalyse (PGF) is effective for estrous induction, but because of variation in body size, 2.5 mg is recommended.

PGF can be used to remedy a mismating, when the doe is bred to the wrong buck. Abortion occurs forty-two to seventy-two hours after injection. If does are aborted during the breeding season, an inter-estrous interval of two to fifteen days can occur. Ovulation does not occur during these short cycles; therefore abortions during the breeding season should be initiated with caution if immediate rebreeding is intended.

PGF can be used to induce parturition (birth) in does. They should not be injected before 144 days and preferably as close to the normal expected birth date as possible. Kidding occurs thirty to thirty-six hours after a single injection of 20 mg of Lutalyse.

Progesterone

Progesterone is a hormone that signals the doe that she is pregnant and helps maintain the pregnancy.

Hormone-impregnated sponges are inserted into the vagina with a special applicator. These sponges are then removed in seventeen days, and the goats generally show heat and can be bred within the next two to four days. Treatment with too high a level of progesterone can decrease fertility.

Syncromate B implants (half the cow dose) can be inserted under the skin of the ear or the underside of the tail. The implants are removed in nine to eleven days, and the does will come into standing estrous within twenty-four hours. Norgestomet ear implants are also cut in half (3 mg) and implanted under the skin. They are left in place for eleven days. Twenty-four hours prior to implant removal, the doe is given 400 IU Pregnant Mare's Serum Gonadotrophin (PMSG) and 50 micrograms of clorostenal. The doe should be in standing estrous forty-three hours later.

Pregnant Mare Serum Gonadotrophin (PMSG)

This hormone can be used after the removal of the progesterone sponges to enhance ovulation. The use of high doses given prior to sponge removal will cause super ovulation and can result in litters. These kids are often aborted, stillborn, or small and weak.

Gonadotrophin Releasing Hormone (GnRH)

This hormone is used in the anestrous period and may induce ovulation, depending on the stage of anestrous. The hormone cocktail (combination of other hormones) used will influence whether heat will occur naturally or must be induced by another hormone, estrogen.

Synthetic Luteinizing Hormone (LH), Human Chorionic Gonadotrophin (HCG)

These hormones will induce ovulation when administered at the first sign of heat. HCG is the most commonly used hormone in this category.

Dexamethasone

This hormone is used to induce parturition, especially if the doe is suffering from pregnancy toxemia. The doe should be no earlier than 144 days. The doe usually kids between 90 and 150 hours after an injection of 16 mg. Dexamethasone as an anti-inflammatory drug should not be used in late pregnancy, as it may cause abortions.

Out of Season Breeding

Melatonin

This hormone has attracted attention lately in the popular press as the solution to aging. Associated with this popularity is considerable controversy over whether this is a drug or just a dietary supplement.

During periods of short daylight, a gland (pineal) located in the brain secretes high levels of melatonin. This hormone is believed to have a stimulatory effect on breeding activity. Experiments have shown that if melatonin is fed to or implanted in sheep for seven to ten weeks, breeding activity occurs outside the normal breeding season. This hormone is not commercially available in Canada.

Altering the Length of Daylight

A producer in Wisconsin exposed his confined does to twenty hours of timed light a day, beginning January 1 and ending March 1. He used banks of 40-watt fluorescent bulbs eight feet long, with one foot of light for each ten to eleven square feet of floor space. By May and June, a large number of these does came in heat and were bred to kid in the fall. The bucks were also light-stimulated. This producer found the milk production to be 15 percent less in the fall-freshened does.

Artificial Insemination (AI)

AI can be used successfully and avoids the necessity of keeping a buck. Most AI units carry frozen semen from various breeds of purebred goats. The bucks all have performance records. Short courses are available to train the producer in AI, or AI technicians are available. To be successful, the goat must be in standing heat; uterine or deep cervical insemination is necessary. These structures must be seen through a vagina scope rather than doing a blind insemination.

Pregnancy Diagnosis

Return to Estrous

During the breeding season, if the doe fails to come back in heat after breeding, one assumes she is pregnant. Good breeding records for each doe are essential to identify pregnant and problem does.

Radiography

Radiographs (X-rays) can be taken as early as day sixty-five in gestation; by day seventy-five a definitive diagnosis can be made. This procedure is not practical or economical for on-farm use. Radiographs may be useful on an individual animal basis; the clinic cost can be up to $250.00.

Progesterone Assay

Blood or milk progesterone concentrations can be used twenty-one to twenty-three days after breeding to detect pregnancy. Open does can be detected with a 95 percent accuracy.

Esterone Assay

Esterone is a hormone produced by the living fetus. If the doe is milking, this test can be done on the milk after fifty days of gestation. Plasma Esterone sulfate can be measured in dry does or does in their first pregnancy. Samples can be sent to a laboratory for testing, or kits can be purchased for home testing.

Ultrasound or Doppler

Ultrasound is used to visualize the fetus through the abdominal wall. A real time ultrasound machine produces an image of the uterus and developing fetus. An experienced operator can detect pregnancy at twenty-five days post-breeding. Fetal numbers can be counted after fifty days of gestation.

Ultrasound machines are expensive. The Doppler ultrasound is less expensive and detects the fetal heart sound through the rectum. Both ultrasound methods require an experienced person to operate and interpret. Cost to the producer will vary. This area of technology is changing rapidly, and the cost may decrease. A combination of transrectal (through the rectum) and transcutanaous (through the skin) after thirty-five days should diagnose most pregnancies. At three months, a single transcutanaous is adequate to detect pregnancy.

Laparotomy

This technique requires a surgical opening of the abdomen and would only be used as a research tool.

Rectoabdominal Palpation

This technique is described in the literature. The doe is tranquilized and placed on her back. A rod is inserted in the rectum, and the fetus is bumped. This procedure is associated with abortion and rectal perforation and has lost its popularity.

Ballottement

At three and a half to four months of gestation, the fetus can be balloted in the right flank of a relaxed doe. Strong-willed does may resist and tighten their abdominal muscles, making palpation difficult.

Summary

To successfully manipulate the reproductive cycle of the doe, the producer must understand the normal cycle and the associated behavior of the doe. The producer is advised to consult with a veterinarian before using any hormones. Just because the neighbor has some leftover is not a good reason to experiment. All hormones, except dexamethasone, are by veterinary prescription only. Since most are human products, accidental injection into the producer may cause problems. If your veterinarian is not familiar with the use of these hormones, you must contact experts in this field. Hormonal manipulation is expensive and should be used for specific reasons (e.g., introduction of a new breed or by producers with experience or money to burn).

Fetal Wastage in Sheep and Goats

The word abortion strikes fear in a producer's heart (at least it did to me). The purpose of this section is to educate you about fetal loss and the steps to take to identify the cause and initiate preventive measures.

Many perils exist along the way to a successful pregnancy. The mother's nutrition and health and the blood (nutrient) flow to the uterus, placenta, and fetus are all key factors. If one or more of these are impaired, early embryonic deaths, abortions, stillbirths, and small, weak, or deformed kids can result.

Early embryonic deaths (one to fifty days post-breeding) occur if the environment in the fallopian tube and/or uterus is altered. Clinically, the only abnormality evident is does coming into heat after missing one or two heat cycles. If you don't know that your buck is breeding, you may not know a problem exists until your does start kidding twenty-one days late, are barren, or kid over a prolonged period. Always have the buck checked if does return to heat during the breeding season. Also, make sure there are adequate bucks to do the job. In dairy cattle, if a very soluble protein or urea is fed in early lactation prior to breeding, ammonia builds up in the fallopian tubes (which move eggs along to get from the ovary to the uterus), and the embryo dies and the doe comes back in heat. Whether this applies to goats is not known. But to avoid any problems do not feed a nonprotein source like urea to goats.

The placenta acts as the border guard between the mother and the developing fetus. Nutrients in the blood pass through or are altered by the placenta to support the growth of the fetus. The placenta may act as a barrier to potential toxins. Some toxins can freely pass through the placenta into the fetus (see Table 16). If problems occur in placental growth or function, its metabolism is disrupted. As a result, the fetus may not receive adequate nutrition, thus impairing its development. Harmful microorganisms can take up residence in the placenta and alter its function. Some of these microorganisms can invade the fetus as well (see Table 17).

Parturition (birth) is initiated by the fetus when the mother's metabolism can no longer support fetal growth. If the placenta's metabolism is impaired, inadequate nutrients are delivered to the fetus. As a result, the fetus may be born too early and not survive. Sudden severe metabolic stress on the doe late in her pregnancy can result in abortions four to five days after the incident. One of the most devastating "abortion storms" I saw in sheep was associated with shearing and a blizzard two days later. The ewes started to abort three to four days later. Unless you are prepared to keep sheared pregnant ewes warm and well fed, I would not advocate shearing as a management procedure.

Some viruses, toxins, and nutrient deficiencies that affect early fetal development can cause mild to severe birth defects (see Tables 16 and 17). Certain nutrient deficiencies late in pregnancy can result in stillborn or weak kids. Some important nutrients that accumulate after a hundred days of pregnancy are vitamin A and E, calcium, phosphorous, iron, iodine, selenium, copper, and zinc. Birth defects can also have a genetic base.

The first abortion is the most critical. If an infection is the cause, early diagnosis may allow for a more successful treatment to avert an "abortion storm". Since 20 to 40 percent of abortions are not diagnosed at necropsy, the fetus and the placenta should be submitted to improve the success rate. Since most abortions are sporadic and do not lead to an abortion storm, early diagnosis can put your mind at ease. Always handle aborted fetal material with caution, as some infectious agents are a human health risk (see Table 17).

Typically, does are not clinically ill prior to or after an abortion. All aborted does should be isolated immediately and the area where she aborted cleaned. The cause of the abortion will dictate whether she should be culled.

Throughout pregnancy, always monitor the nutritional status of your does. Are they too thin or too fat? Any dietary changes should be made gradually. Sudden dietary changes near term should be avoided, as they may precipitate pregnancy toxemia.

When you purchase replacement animals, investigate their backgrounds carefully. Avoid buying someone else's problems; they could haunt you for years.

Once the cause of the abortion is determined, consult your veterinarian regarding the significance of the necropsy findings, appropriate treatment, and potential preventive procedures (e.g., vaccines available and their limitations).

Table 16: Environmental and Dietary Factors that Can Result in Fetal Wastage.

Factor	Stage of Pregnancy*	Impact on Fetus	Solution
Does fed 30 percent below requirements	30 to 90 days	Increased mortality Survival of one twin	Early evaluation of diet Worm does Monitor body condition
Impaired placental growth	> 100 days	Intrauterine growth retardation Small, weak kids	Adequate nutrition during entire pregnancy
Heat stress	early pregnancy	One degree centigrade increase in body temperature decreases uterine blood flow 20 percent to 30 percent Sporadic malformations	During breeding season, provide shelter from heat
Metabolic stress e.g., shearing in late pregnancy	> 140 days	Abortions 3 to 4 days after stress	Prepare for cold stress
Deficient Vitamin A	> 100 days	Abortion, retained placenta Weak, blind lambs	Evaluate diet Supplement Vitamin A

Deficient Vitamin E/ selenium (Se)	> 140 days weathered large round bales are deficient in all vitamins	Born with white muscle disease	Evaluate diet Supplement Vitamin E/Se Inject Vit E/Se 20 days prepartum (avoid stress)
Deficient iodine**	> 50 days	Early embryonic death Abortion, still born Goiter, naked Prolonged gestation	Loose cobalt/iod-ized salt Free choice or add-ed to diet
Hereditary goiter	Seen in a strain of Dutch goats and in Boer goats	Autosomal recessive trait Clinically the Boer goats are normal; they have a large thyroid In the Dutch breed, the kids are weak and grow slowly	Both parents carry the recessive gene Cull both
Deficient copper	> 100 days	Kids born unable to walk properly (ataxia)	Evaluate diet for Cu and nutrients that interact with Cu Avoid Cu toxicity
Deficient manga-nese	> 50 days	Abnormal bone de-velopment Ataxia Contracted tendons	Does diet adjusted to supply 40 mg Mn/kg diet
Asralagalus (Loco-weed)	0 to 154 days	Heart failure Contracted tendons	Remove toxic weed from diet
Veratrum californi-cum	14 days 19 to 21 days 28 to 33 days	Kids born with one eye Prenatal death Limb defects, flat trachea	Remove from diet
Conium maculatum (Poison hemlock)	30 to 60 days	Arthrogryphosis (crook-ed kids)	Remove from diet
Nicotiana glauca (Tree tobacco)	30 to 60 days	Cleft palate, skeletal defects	Remove from diet

* Stage of pregnancy when fetus is most vulnerable

** Goiter should not be confused with a large thymus, which presents as two equal sized swellings in the region of the thyroid. These become noticeable at two weeks of age and regress at about four months.

Table 17: Microorganisms That Can Cause "Abortion Storms" or Sporadic Abortions

	Incidence	Organism	Source	Stage of Pregnancy	Lesions	Treatment/ Prevention
Enzootic Abortion*	47 percent of diagnosed abortions Abortion storms Endemic flocks 5 to 10 percent annual abortions	*Chlamydia psittaci* (ovine)	Feces of carrier does Fetal fluids from abortions Aborted does shed up to 20 days	80 to 110 days Invades placenta	Placentitis (infection in the placenta) Abortion 2 to 3 weeks premature Aborted fetus fresh	Long acting tetracyclines 20mg/kg BW 10 to 14 d interval (only useful if placental lesions are mild; once abortions occur too late) Vaccine available
Toxoplasma*	50 percent of kids can be aborted	Protozoal parasite	Infected cat feces Survive in soil several months	Mid-gestation	Embryonic death Barren Placentitis Abortion, mummified fetus, weak kids	Aborted does immune Decrease cat population 10 to 20mg monensin/ head/d as a prevention in problem areas
Vibriosis	Susceptible flocks 70 percent abortions Endemic flocks 5 to10 percent abortions	Camphylo. jejuni or fetus	Carrier animals Fetal fluids Shed several weeks postpartum	Incubation period 8 to 60 days	Fresh fetus stillborn or weak	Aborted does immune Vaccine available
Border Disease (Hairy Shaker)	Associated with intensive management	BVD virus	Carrier sheep and cattle	Related to clinical signs	Barren ewes Abortion, stillborn Lambs with tremors Hairy fleece	Avoid carriers
Brucellosis	1 to 5 percent abortions Rare	B. ovis			Infertility in rams	Blood test Cull infected goats

Salmonella*	Rare but devastating	S. arizona, typhimur., dublin	Carriers Survive long time in dry manure		Ill prior to abortion	Cull infected animals
Neospora caninum	Recently identified	Protozoal parasite	Carnivore host	90 days	Abort rotten fetus 25 days post-infection Placentitis	NA
Q Fever*	Rare, related to stress	Coxiella burnettei	Tick borne? Shed in placenta	NA	Placentitis Rotten fetus	Control stress
Cache Valley Virus	Rare	Virus	Gnats, mosquitoes	27 to 45 d	Crooked kids	Blood test
Akabani Virus	Rare	Virus	NA	30 to 36 d	Birth defects	Blood test
Listeriosis	Rare Associated with silage	L .monocytogenes	Mouse feces contaminate feed (silage)	140 d	Abort 5 to 10 d after infection Fetus rotten Sick with metritis (infection in the uterus)	Avoid feed contamination
Fusobacter. nucleatum	Sporadic	*F. .nucleatum*	?	?	Placentitis Fetal pneumonia	?

* Organisms pose a human health risk; handle fetal material with care.

Chapter 5

Common Diseases

Introduction

How do I determine if my goat is sick? No matter how many goats you have, you should make a daily check on all of them. To momentarily look at them while feeding may not be enough. Sick animals may temporarily respond to the excitement around them. Watch them for a few minutes after feeding (you deserve the break anyway). This is the best time to check for a boss goat and to observe how the group interacts. This is a good time to just observe normal animals so you can detect early changes. Look for a goat that backs away.

Know the history of your animals. The following information is essential when you call your veterinarian:

- Age
- Sex
- Stage of production
- Vaccination and deworming program
- Current and past history (records)
- Length of time in herd
- Hand-raised or purchased
- Earliest symptoms

Examination of the Goat

Examination from a Distance

It is important to stand back and observe your goats. Get to know how they look and act when they are normal so that you will be able to detect changes if they get sick. While you observe from a distance, answer the following questions:

- Is there a change from her normal **behavior or attitude**? Is she showing pain?
- Is there a change in her **body condition**? Is she losing weight?
- Are there any obvious **swellings** of joints or under the skin?
- Does she stand with her **back** humped up?
- When she relaxes does she **chew her cud**? (During a quiet time 50 percent of the goats should be chewing their cud.)
- Is her **manure** normal?
- Does she have difficulty **passing manure or urine**?
- Is her **breathing pattern** normal (both standing and lying down), or is the rate increased (normal is 21 to 27 breaths per minute)? Does she have trouble breathing in or out?
- Is the contour of her abdomen normal?
- Is her **hair coat** dry, dull, or rough? Is her winter coat slow to shed? Are there patches where hair is missing?

Close up examination

If you feel there is a problem, you should catch the ailing goat and have a closer look. Answer the following questions:

General Examination

- Was she easier to catch than normal? Does she seem to walk in a circle or stagger?
- Does she have any lumps under the skin? Does the hair and skin look normal?
- Is she "hide bound" (dehydrated?) Are her eyes sunken?
- Is she emaciated?
- Has she been rubbing or scratching herself?

Rectal Temperature

- Is her temperature normal (102.5 degrees F [39.4 degrees C] to 103 degrees F [39.6 degrees C])?
- If the weather is hot and you had to chase her, wait for a while before checking her temperature, or take her temperature again after she calms down a while. Normal or below temperatures does not mean she is normal.

Nervous System

- If she is down can she get up easily? Is she paralyzed in any leg? If you pinch the skin between her toes, can she feel it? If excited, does she go into convulsions (on her side with her head back and legs stiff)?
- Does she appear blind? When in a dark place, do her pupils dilate? If she is outside in the bright sun or if a bright light is shone in her eyes, do her pupils constrict? Is there any rapid eyeball movement up or down?

Head and Eyes

- Are her eyes dull or cloudy?
- Is she alert?
- Is the face normal? Does one ear droop? Any lumps?
- Evidence of drooling? Cud in cheek?
- Check along the outside of the cheek where the teeth are to see if there are any painful areas or abnormalities.

Mouth

- Examine the gums and front teeth for abnormalities.
- Her gums should be pink and wet.
    - Note: The mucous membranes inside the vagina and under the eye lid (conjunctiva) should be the same color.
- If blue or muddy-colored, press your thumb into the gum and then remove; the color will return immediately if circulation is normal. If the blanched part takes longer than one second to refill, she is going into shock
- Does her breath smell unusual (e.g., acetone smell = ketosis; ammonia smell = kidney or liver disease)?

- Does she have bad teeth? Are there fox tails or barley awns lodged in her mouth?
- Is her tongue normal? Can she move it?

Note: her back teeth are very sharp and difficult to examine orally.

Udder

- Is her udder soft and pliable?
- Are the teats normal?
- Is the milk normal?

Respiratory System

- Does she cough or have a nasal discharge? Is the discharge from both nostrils or just one?
- Is air coming from both nostrils?
- Use a mirror look for condensation of breath from each nostril.
- Does she appear to be having trouble breathing in or breathing out?
- If you squeeze her wind pipe gently does she cough?

Digestive System

- Is the contour of the abdomen abnormal? Is the abnormal area soft and gas-filled or firm? Is this area painful?
- Is her manure normal, loose, dry and covered with mucous, or scant? Is her manure a normal color? Is there blood mixed with the manure, or on the surface? Fresh blood is red from lower intestine; digested blood is black (tar-like) from upper intestine and stomach.

Use of a Stethoscope

You do not need to be a veterinarian to use a stethoscope. Practice listening in the following areas in normal goats so that you become familiar with the sounds. You can also listen to the sounds in your own chest and abdomen.

Wind pipe

Become familiar with the sound of a normal breath in and a normal breath out.

Left side under the elbow

Listen for the normal heart beat.

Left and right side under the elbow

Listen for the air moving in and out of the lungs.

Left and right side over the rib area

Listen for the air moving into and out of the lungs.

Left side over the flank

Listen for the normal sound as the rumen moves, which sounds a little like waves as they wash up on a beach.

Right side over the flank

Listen for the normal gurgling sounds and movements of the small intestine

Group Problems

Occasionally you may encounter a problem in a group of goats. The problem may be as simple as a mild indigestion or as complex as sudden deaths in a group of goats. Group problems can be devastating emotionally and economically. A change in feed or feeding practices or moving goats into a new environment can lead to issues; malicious or accidental and inappropriate feed mixing or drug dosages should be considered when a poisoning is suspected. Digestive upsets can be attributed to sudden change in water or lack of water, sudden change in feed or feed bunk management, or parasites. Contagious diseases such as pneumonia or coccidiosis are related to lapses in environmental management. Nutritionally related problems will appear in a group of goats.

PHYSICAL EXAMINATION SHEET

As you examine your goat, record you findings. These findings can be useful when talking to your veterinarian and can be attached to the goat's records.

Date: ______________________________ Goat's ID: ____________________

History: ____

____

____

____

____

____

Rectal Temperature: ____________________ Respiratory rate: _______________

Symptoms: ____

____

____

____

Distance Examination and General Appearance: ______________________________

____

____

____

Head and Eyes: ____

____

____

____

Mouth and Tongue: ____

____

____

Nervous System: ____

____

____

Digestive System: ____

____

____

____

____

____

____

____

____

____

____

____

____

The answer may be in the feces

The nature of a goat's feces is not often volunteered as part of your history when you call a veterinarian. Yet examining the feces will help you evaluate the problem in either a single goat or in a group of goats. The characteristics of the feces should not be solely used in making a clinical diagnosis but can be helpful if combined with the history, age, and other clinical signs in diagnosing a goat's problem.

A goat is very efficient at grinding its ration during rumination and removing the water from the feces as the ingesta move through the large intestine. Thus the feces passed are typically dry, hard pellets with hardly any recognizable dietary components, such as whole grain. Occasionally the feces may be coated with a thin layer of mucus. Unless the goat is showing clinical signs of an illness, this is likely normal. The following is an attempt to help you interpret the nature of the feces in relation to the age and clinical signs the goat(s) is (are) showing.

Newborn to Four or Five Days of Age

The first feces a newborn kid passes is called meconium. It is very sticky and dark yellow to brown in color. The kid may pass the meconium while in the uterus, especially if the birth was prolonged or difficult. When this occurs the hair will be stained yellow. Because of the sticky nature of meconium, it may stick to the hair around the anus and can actually block the passage of feces, making the kid uncomfortable and appearing constipated. If the doe does not remove the meconium, you may have to use warm water to soften the meconium or clip the hair away to clear the anus.

Colostrum, along with its important role in providing immunoglobulins (maternal antibodies) to protect the newborn from diseases, is mildly laxative and assists the kid in passing the meconium. Thus the early feces passed by the kid will be soft.

If no feces are passed in the first few hours of nursing, the kid, who initially appeared bright and was up nursing, may become slightly bloated, uncomfortable, and depressed. This kid may have a congenital problem called segmental aplasia, where part of the intestinal tract failed to develop. To check for this, insert a thermometer gently into the rectum. If you cannot insert it very far and when removed there is no meconium sticking to it, then the developmental defect is close to the anus. If you can insert the thermometer but it comes out clean, then the defect can be anywhere. Call your veterinarian if you suspect this; in most cases euthanasia is recommended.

If the kids are put on milk replacer after nursing colostrum, they may develop nutritional scours. Twice daily feedings with milk replacer to meet the kid's nutritional requirements does not represent a natural nursing pattern and is often initially difficult for the intestinal tract to adapt to.

During this period kids start to nibble on the straw, feed, or feces present in the environment. This behavior is necessary because the kid picks up bacteria and microorganisms that help in the normal development of the gastrointestinal tract. Unfortunately, the kid may also pick up harmful microorganisms such as coccidia, harmful E. coli and salmonella, viruses, intestinal parasites, or the bacteria associated with Johne's disease.

Kids from Birth to Six Months of Age

Scours related to bacteria or viruses are not as common in goats as they are in cattle. They are often associated with stress and a build-up of these microorganisms in the environment prior to kidding. The feces can be moist and soft or very watery and frequently voided. Blood and fibrin casts (looks like intestinal lining surrounding a jelly-like substance) indicate that the harmful microorganisms are causing the destruction of the cells lining the intestinal tract, exposing and injuring small blood vessels. When this happens it may take up to seven days for the wall of the intestine to heal. Some of the kids will remain nursing, but in many cases dehydration, acidosis, an electrolyte imbalance, and a low blood sugar cause the kid to become weak and stop nursing and its mouth to become dry and cold. These kids require electrolytes with dextrose specially formulated for oral, parenteral (subcutaneous—under the skin), intravenous, or Intraperitional (into the abdominal cavity) use. The use of antibiotics is a hotly debated subject. I personally do not recommend them, as they not only can kill the bacteria causing the problem but also kill all the susceptible good bacteria necessary for a healthy gut, which can prolong the scours. To help restore the normal bacteria in the gut, plain yogurt (e.g., Activia) can be fed.

Coccidia and intestinal parasites usually cause scours in kids over twenty-one days of age, as these organisms must first complete their life cycle. Coccidia can cause a mild scours or a severe bloody scours with fibrin casts. Severe scours may be associated with straining, and on rare occasions the kid may prolapse its rectum. If the blood loss is severe the kid will be anaemic. Both conditions can be treated with electrolytes, but successful treatment will depend on whether coccidia and/or eggs of the intestinal worms are found on a microscopic fecal examination. Once the diagnosis is established, appropriate treatment and preventive measures can be taken. Scours associated with coccidiosis is more common in kids under six months of age and often follows severe stress. Intestinal parasites can occur at any age; scours is dependent on the numbers present.

Moving white, rice-like segments (individually or linked together) can be seen in the feces of kids on pasture. These are tapeworm segments and are not considered a problem except in severe cases. They may indicate a more serious intestinal parasite problem, which can only be diagnosed through a microscopic fecal examination.

Six Months and Older

The following are some changes in feces that may help you come to a diagnosis of a clinical problem:

- Normal feces coated in mucus
    - May be normal
    - Can indicate a fever or dehydration
    - Not eating or poor quality diet
    - Lack of water or adequate water
    - Advanced pregnancy
- Soft feces (cow-pie like)
    - Change in diet from winter forage to pasture or related to increased grain in the diet. Can also be related to a change in water
    - Intestinal parasite problem
    - Late stages of Johne's disease

   - Sudden stressors
   - Associated with certain medications, such as Ketol for pregnancy toxemia or antibiotics
- No feces
   - Associated with colic or straining; may indicate an intestinal blockage
   - May just pass a little foul-smelling mucus mixed with blood

An Overview of Some Problems

Although a long list of diseases and problems can be developed (to sell books, confuse veterinary students, and make owners paranoid), once a kid has survived to weaning, the neighbors dogs are confined, and the last coyote/wolf has left, your goat will likely live a productive life.

The following tables outline some of the more common problems I have seen in goats over the years. Fortunately goats are hardy, and if fed and managed well, they will reward you with years of production and low veterinary bills. If goats do get sick, depending on the cause, they tend to get very sick. They are sensitive to pain and, when under extreme stress, can go into shock and die.

Congenital Diseases

Kids can be born with defects that developed while in the uterus. The causes of these congenital diseases are difficult to determine because the damage often occurs during organ development early in pregnancy. The common causes are dietary, inherited, teratogens (environmental or plant toxins that damage the developing fetus), and viruses.

Table 18: Examples of Some Congenital Diseases

Body System	Disease	Cause	Lesion	Key Features	Prognosis	Treatment/ Prevention
Brain	Hydrocephalus	Multiple	Fluid builds up inside brain	Blindness convulsions Uncoordinated	Die shortly after birth	No treatment
	Cerebellar Hypoplasia	BVD virus (hairy shakers)* Unknown	Part of brain that controls movement is not developed	Uncoordinated Stands with feet far apart Head tremors when tries to nurse Kids weak at birth and they shake	If severe, kids die If mild, with special care kids can survive but do poorly	None
	Enzootic Ataxia (swayback)	Doe's diet deficient in copper	The membrane around the nerves in the brain not developed Nerves short circuits	Uncoordinated in hind end, recumbent Difficulty nursing Can occur in older kids	Poor	Ensure doe's diet adequate in copper (see nutrition section)
	Caprine B-mannosidosis	Genetic; recessive carried by both the parents	Lack of an enzyme build-up of abnormal sugars in the brain	Whole body tremors Paralysis	Poor	Cull both the buck and doe
Lumps under the neck skin	Excess thymic tissue	Genetic (Nubians)	Excess thymic tissue	Kids bright and alert Lumps under the skin in the neck	Regress in 7 months	Not that serious
	Goiter	Iodine deficiency Some breeds (Boer) have a genetic predisposition	Iodine deficiency	Severely weak or stillborn, may lack wool, large thyroid glands Mild, small goiter	Poor in severe cases	Mild cases 5 drops of Lugol's Iodine in one oz of water by mouth per day for 5 days Free choice loose cobalt iodized salt all year round
Reproductive	Intersexes	Genetic, related to the polled trait	Females sterile, have rudimentary male sex organs	Clitoris is large	Infertility	Chance you take when breeding for polled animals
	Freemartins	Placental fusion between male and female	Uterus and ovaries do not develop	None	Female is infertile	Very rare Common in cattle
Skin	Wattles	Dominant gene				Does with wattles have 0.13 more kids than those without

Diseases of Neonates to Puberty

This period has the highest death rate related to many factors. Most are related to management, but some are beyond your control.

Table 19: Diseases: Neonate to Puberty

Problem	Age (Day)	Cause	Clinical Signs	Prognosis	Treatment	Prevention
Rag Doll Floppy Kid Syndrome	3 to 10 Late in kidding season	Abnormal levels of acid in the blood Exact cause unknown, suspect abnormal bacteria build-up in intestine	Depressed Weak, limp Excess fluid in abdomen	Good if treated early Death occurs in 10 percent to 50 percent Some recover spontaneously	If severe— IV Na Bicarbonate Mild signs oral bicarbonate	Avoid using oral antibiotics as a preventive measure for scours immediately after birth
Navel Infection	1 to 14	Wet, dirty environment Inadequate colostrum Poor or inexperienced mother	Depressed, fever Navel wet painful Reluctant to walk Enlarged joints	Fair if joints, liver or brain not infected Can cause an abscess in the spinal cord and paralysis	Penicillin 20,000IU/kg BW	Prepare kidding area early Ensure adequate colostrum Dip or spray the navel in an iodine solution + monitor daily until cord is dry Provide a well bedded creep area
Joint Ill	6 to 21 days	Complication of a navel infection	Depressed Lame Enlarged, swollen joints (knee, hock most common)	Guarded; since the infection is carried in the blood, other organs may be involved (e.g., heart valves)	If only one joint involved, it can be opened and drained; call your veterinarian	As above
Coccidiosis	Life cycle of parasite is 14 to 21 days Organism shed in adult's feces	A small parasite that completes its life cycle in the cells lining the intestine It breaks out of the cell when mature, causing scours and bleeding	Depends on the number of parasites Sudden death Depression, abdominal pain Straining to pass feces Scours often bloody, dark brown, with fetid odor Lose weight, dehydrated May become anaemic	Good to guarded Some kids will do poorly after In most will clear up in 4 to 7 days when the intestine heals Partial immunity develops	Oral electrolytes Sulpha drugs are only effective before the organism breaks out of the cell AD injection	Keep bedding clean and dry Eliminate sources of stagnant water Coccidiostat* in the creep ration—e.g., Deccox 0.5 mg/kg BW; Rumensin 11 mg/kg of feed; Lasalosid Amprolium in the water * Note: coccidiostat must be prescribed by your veterinarian

Orf (sore mouth)	>2 weeks	Virus	Cold sore-like lesions on gums, lips, around the eyes May be reluctant to nurse	Good, resolves in 1 to 4 weeks	None, leave alone, as orf can infect humans	Vaccine not recommended as it is a live virus
Pneumonia	If under 2 weeks	Viral/bacterial Primarily management: crowded, poorly ventilated area, not cleaned or well bedded	Depressed Increased rate but shallow breathing, cough Fever	Good if management changes	Antibiotics if severe Of little value if housing and ventilation not improved	Get down on your hands and knees at kid level to see what the air is like Improve housing conditions Improve ventilation
Enterotoxemia	Weaned	Bacteria Clostridium perfringens Type D High grain diets or lush pastures	Sudden death Can be acute or chronic, where diarrhea is common	Difficult to treat	If alive treat with penicillin	Avoid sudden feed changes, especially from poor to rich Vaccine not as effective as in sheep
Caprine Arthritis Encephalitis (CAE)	Any age	Viral Spread through colostrum & milk (1 ml is enough)	Kids: Gradual onset of hind end paralysis; they remain alert Adults: Swollen joints mildly painful Chronic interstitial pneumonia Hard Udder	Poor	No treatment Infection for life 10 percent of kids from infected dams develop antibodies independent of preventive procedures	Confirm through blood tests AGID or ELISA, once every 6 months Remove from mother before nursing Feed pasturized colostrum or from a clean doe ·Raise as orphan Maintain two groups, one negative, one infected
White Muscle Disease	From birth to 6 months of age	Nutritional deficiency of Se and or Vitamin E Large round bales of hay that have been down for more than two days will be low in Vitamin E.	Kids are stiff and reluctant to move Muscles are hard and painful If the heart is involved, heart failure or sudden death If the tongue is involved, kids are unable to nurse Pneumonia-like signs are evident if the diaphragm is affected	Prognosis is good if caught early	Oral or injectable Selenium / vitamin E	Selenium / vitamin E supplements Caution: Avoid feeding multiple supplements that contain added Se. Over- supplementation with Se can cause a toxicity.

Table 20: Some Examples of Disease Problems in Kids from Weaning to Puberty

Disease	Age	Causes	Clinical Signs	Prognosis	Treatment	Prevention
Pneumonia	Any age May follow stress	Environmental stress Poor ventilation Crowding Virus/bacteria/ lung worms	Fever Depressed Off feed Rapid breath-ing Cough	Good if man-agement of the environment improves	Antibiotics for bac-terial problems Select appropriate dewormer for lung worm	Improve en-vironmental conditions Routine parasite control
Polioencephalomal-acia	2 months to 2 years	Thiamine de-ficiency or an inhibition of thiamine activ-ity related to high grain diets; amprolium (>8.2 gm/kg body weight for >6 weeks other unknown factors	Nervous, ataxic, convul-sions, blind-ness	Good if treat-ed early Blindness can last up to 6 weeks	Thiamine IV 10 mg/kg BW every 6 hours for 24 hours. Thiamine can be injected into the muscle	Look for cause
Vestibular Disease	Any age	Inner ear infec-tion, secondary to ear mites or bacteria	Sudden onset Head tilt, loss of balance	Good if treat-ed early	Antibiotics e.g., penicillin Check for mites	None
Tetanus	Any age	Bacteria: can occur fol-lowing: Castration with elastic rings Deep puncture wounds Traumatic kid-ding	Become stiff, unable to move If excited, be-come stiff and fall over Mild bloat	Very poor	Open and drain site of wound High doses of peni-cillin	Vaccinate does before kidding and the kids at 6 to 8 weeks. Repeat in 3 weeks and then at one year

Chronic Weight Loss in Does

The following problems are common causes of weight loss in adult goats. Some causes are briefly described in the section on blood tests.

Caseous Lymphadenitis

+ **Human risk Caution:** Handle the animal with care, as this disease can occur in humans as a chronic lymphadenitis.

Cause: *Actinobacillosis pseudotuberculis* is an organism that causes abscesses in the lymph nodes of the body. The bacteria can enter through cuts in the skin, and feed and water can be contaminated by draining abscesses.

***Clinical signs*:** Signs are associated with the site of the abscesses. If the disease is only in the lymph nodes under the skin, these nodes will be swollen. The most common site is around the head and upper neck. If the abscesses are internal and if the lungs are involved, there may be difficulty breathing and weight loss. If the liver and kidney are involved, chronic weight loss may be the only sign.

Prognosis: Poor. You should cull the animal. If it is slaughtered, it may be condemned because of abscesses and emaciation.

Treatment: There is no treatment, and the vaccine does not eliminate the abscesses; it only reduces the number. Occasionally, enlarged abscesses are surgically removed. Success depends on the location of the abscess and whether or not it is the only one.

Prevention: Purchase replacement animals from producers willing to show you all their animals and their records. Isolate any goats with external abscesses that are draining. If she is not culled, keep in isolation until abscess dries up. The vaccine Glandvac 3 is no longer approved for goats. Do not lance lumps on your goats.

Caution: I have seen animals bleed to death because a hematoma (broken blood vessel) was lanced or cut through. If the lump is an abscess, it may not yet have come to a head (point). If lanced prematurely, a severe local inflammatory reaction may occur. My philosophy is to leave the lump alone and let nature take its course. Isolate goat until the abscess heals.

Johne's Disease

***Cause*:** *Mycobacterium paratuberculosis* can survive in the external environment for up two years. The bacteria invade cells lining the intestines and impair the ability of the cells to absorb nutrients. The infection usually becomes established when kids are under four months of age by exposure to the organism shed in the adult's feces. The infection can be spread from cattle or sheep, but the goat is most sensitive.

***Clinical signs*:** This disease often follows a period of stress (such as kidding). Animals lose weight despite a good appetite. The feces are usually normal, but intermittent loose feces may be seen.

***Prognosis*:** Poor. Early culling is advised, as the organism is shed in the feces.

Treatment: There is no treatment. Blood tests may help to confirm the diagnosis, but they are not always reliable. They do not always identify all the infected animals, nor do they indicate if the goat is shedding the organism in its feces. Occasionally, surgery is done to biopsy the intestine and abdominal lymph nodes, and the samples are sent for culture and special stains.

Prevention: The organism is shed in the feces of the clinically ill and carriers. There is no reliable, rapid, or inexpensive test to identify carriers. Remove kids from mother immediately after birth. Feed colostrum that has been pasteurized or is from an unaffected doe. Keep the kidding area clean and dry. Disinfectants (cresylic compounds, diluted 1:64, and orthophenylphenate, diluted 1:200) can eliminate the organism.

Chronic Pneumonia

***Cause*:** The most common cause is viral (chronic progressive pneumonia). Lungworms and chronic bacterial pneumonia must be included.

***Clinical signs*:** The clinical signs are weight loss, a chronic cough, and increased respiratory rate if stressed.

***Prognosis*:** The prognosis is poor unless the problem is related to lungworms (based on a special faecal examination by your veterinarian).

Treatment: The only condition that responds to treatment is lungworms, if treated with the correct dewormer (see section on drugs approved for sheep and goats).

Prevention: Control of viral pneumonia requires that, at birth, kids be taken away before nursing, fed pasteurized colostrum, and raised as orphans (see CAE, p. 70).

Scrapie [21][22]

Scrapie is controlled by federal veterinarians. If scrapie is suspected, they must be notified.
Note: Health of Animal's protocol must be followed.
Scrapie has gained notoriety lately because of Mad Cow Disease in Europe and North America.

Cause: Goats are susceptible to this disease, but actual clinical cases are rare. It can be transmitted from sheep at lambing time. The cause is a protein (prion) smaller than a virus that invades the brain, resulting in slow degeneration. Clinical progression is slow.

***Clinical signs*:** The doe continues to eat well but loses weight. She is very itchy and rubs against things (look for lice).

***Prognosis*:** Prognosis is grave (literally).

Treatment: There is no treatment.

Prevention: Avoid purchasing replacement stock from unknown sources. Separate sheep and goats prior to and at parturition and up to two weeks after.

Malnutrition

Cause: Malnutrition can be caused by poor quality feed, inadequate feeder space, inadequate water intake, poor teeth or no teeth, mixing too many age groups, or problems with dominant versus submissive goats. One cause often overlooked is the drain on body reserves associated with lactation. All these factors are complicated by lack of shelter and bedding and poor weather conditions.

***Clinical signs*:** The main clinical sign is weight loss. Examination of the mouth will rule out teeth. Molars are difficult to examine, and goat's teeth are very sharp. I examine the back teeth from outside the mouth along the cheek. I look for any swelling and press on the molars to check for pain or missing teeth.

***Prognosis*:** The prognosis depends on the cause and the willingness to correct the management problems.

Treatment: Feed testing and ration evaluation are important. Check the water supply and systems used to deliver the water. Remove the dominant goat.

Prevention: Feeders should allow for 1.4 to 2 ft per goat. If kept in a dry lot, allow 30 to 100 square feet per doe.

Problems Associated with the Pre- and Postpartum Period

Angular Limb Deformity

Cause: Many causes have been proposed, but investigations have not reached any firm conclusions. I believe that this may be related to joint ligament laxity and not associated with abnormalities in the bones.

***Clinical signs*:** This condition occurs in obese does (BCS=5) due to kid before they are twelve months old. The main clinical sign is a weakness in the lower leg joints, causing the limbs to be crooked. This may involve only the front or hind limbs or all limbs. The doe has difficulty walking.

***Prognosis*:** The deformity gradually clears up after kidding.

Treatment: You can minimize the changes by supporting the limbs. Do not put her on a diet at this time, but investigate her diet to make sure adequate and balanced nutrients are present.

Prevention: Monitor the body condition of young pregnant does and avoid over-feeding.

Pregnancy Toxemia

Cause: This is associated with multiple fetuses and the last third of pregnancy, when protein and energy demands are high. It is seen in both over-conditioned and thin does. It usually follows periods of nutritional, metabolic, and environmental stress.

***Clinical signs*:** The doe goes off her feed, becomes depressed, staggers, and eventually goes down. There is a strong acetone smell to her breath. Urine will test positive for ketones (special ketone test tablets can be obtained from your veterinarian).

***Prognosis*:** They can die if not treated quickly.

Treatment: The doe must be monitored closely, as she can develop severe acidosis quickly (acid build-up in the blood). In early or mild cases, hand-feed grain and 60 cc of propylene glycol orally twice a day. Administer orally one tablespoon of baking soda dissolved in warm water, two to three times a day. If not too severe, parturition (kidding) can be induced (see the section on reproduction). Call your veterinarian for severe cases, as the condition may require intravenous fluids containing 5 percent dextrose and bicarbonate. If severe, a caesarean section should be done quickly to salvage the kids.

Prevention: Avoid extremes in body condition. Ensure that the diet is adequate throughout the pregnancy. Reduce stress for does with multiple fetuses.

Ketosis

Cause: A milder form of pregnancy toxemia, ketosis is associated with a lactation drain rather than a fetal drain. It occurs early in lactation, when dietary intake does not meet the demands for milk production and body reserves of fat are mobilized to supply this demand. Ketone bodies are produced and are utilized for energy by some tissue, so that glucose is available for tissue that can only use glucose. A low level of ketones is normal.

***Clinical signs*:** These are not very specific. She may be weak and have a depressed appetite. Her breath will smell of ketones, and her urine will test positive.

Prognosis: Good

Treatment: 60 cc of oral propylene glycol (Ketol) should be given twice a day. If slow to respond to oral treatment, 50 percent dextrose can be given intravenously (1.5cc/kg BW).

Prevention: Avoid extremes in body condition. Ensure that the diet is adequate throughout the pregnancy. Reduce stress for does with multiple fetuses.

Dystocia (Difficult Birth)

Cause: Kids should be born within one and a half hours of active labor. Problems in this area are not common (thank goodness). However, abnormalities can occur in birthing, and there are several possible causes.

- The neonate is too large to be delivered.
- The kid is in an abnormal position or there is more than one kid in the birth canal.
- There is a failure of the cervix to dilate, torsion of the uterus and ring-womb (often the owner notices the doe starts to kid but then seems to stop). Failure of the cervix to dilate may be related to lack of exercise, but the exact cause is unknown. Unfortunately it can be a herd problem.

Treatment: Treatment is dependent upon the cause of the difficulties.

- If the neonate is too large to be delivered, a caesarean section will be necessary.
- If the kid is in an abnormal position, such as the head back, front leg(s) back, coming backward, breech (only tail and hind end present), identify position of the kid (kids), correct the position, and deliver. Most how-to books have illustrations of the abnormal positions.
- If the cervix fails to dilate, the remedy is usually a caesarean. If detected early, prostaglandins can be used to relax the cervix.

Prevention: Cull problem animals, especially those that required a caesarean because of size or failure of the cervix to dilate.

The Mammary Gland

Mastitis

Cause: Mastitis is an infection via the streak canal (teat end), which can occur with improper milking and hygiene or self-nursing. It can be caused by a number of different organisms. If in doubt, contact your veterinarian for a diagnosis and treatment. Some of the organisms involved are:

- *Haemolytic staph aureus*: Severity depends on the immune status and stage of lactation. Clinical signs vary from subclinical to systemically ill with an enlarged inflamed udder, which may become gangrenous (cold and black) and slough.
- *Non haemolytic staph*: There is an increased somatic cell count and decreased production. Diagnosis is based on a culture of the milk.
- *Streptococcal* mastitis: In this chronic condition, abscesses develop in the udder.
- *Pseudomonas*: This can result in a subclinical condition or can cause an acute purulent

mastitis progressing to gangrene.

- *Pasteurella* mastitis—*P. hemolytica*: The gland swells, and secretions are bloody with clots. It may be acquired from the nursing kids.
- *Actinobacillus pyogenes*: This causes slight swelling and a secretion that is watery and straw-colored. The condition declines in five days. There is some scarring of the mammary tissue.
- *Mycoplasma capricolum:* This is associated with arthritis, conjunctivitis, and abortion.

Note: Some of the antibiotics used to treat mastitis are not approved for goats or the dosage necessary is greater than the label dose. In these cases the withdrawal period (timeframe for withholding the milk for human consumption) is difficult to predict. You can check the milk for antibiotics with the Delvo P test. Some dairy farmers have the test kit. If you are concerned, contact a local dairy farmer or a dairy producer's outlet for information on the test.

Hard Udder

At kidding the udder is enlarged and hard, no milk secretion is evident, and the doe is not ill. ***Always cull the doe.*** This has been associated with the CAE virus; others suggest a calcium deficiency. The type of treatment will vary depending on the cause, severity, and prognosis for a successful outcome. Milk out the gland as many times a day as possible. Oxytocin may help. Hot packs and mild liniments may help as well. Mastitis ointments may help, but milk withdrawal times are not known. ***Do not use the tip of the cattle ointment container, as it is too large***. A tom cat catheter is ideal (see your veterinarian).

Dry Goat Treatment

Infusion of dry cow ointment (1/2 tube) after the last milking can decrease the incidence of subclinical mastitis in the next lactation.

Anomalies

- Supernumerary (extra) teats: Remove small extra teats shortly after birth.
- Abnormal teats: Sell the doe for meat.
- Small teat sphincter: This is congenital or traumatic in origin. You can try to enlarge it with a 20-gauge needle (instruments used for cattle teats are too large).
- Witch's milk: Enlargement of a kid's udder shortly after birth.
- Gyneomastia: Enlargement of the rudimentary teats in bucks and secretion of a milk-like substance, associated with a hormonal imbalance and infertility.
- Precocious udder: Unbred doe can develop a lactating udder.

Disease of the Skin of the Udder

- Contagious ecthyma (orf): Viral infection in kids (cold sore-like lesions on lips and face) can spread to the doe's udder. + **Human risk** (can infect humans)
- Goat pox is not of concern in Canada.
- *Caprine papillomatosis*: wart virus
- Chorioptic and sarcoptic mange mites: These are present in other areas of the body.
- Reaction to biting flies.
- White-skinned goats are at risk of a sunburned udder.
- Poor udder conformation can lead to trauma.

- *Staph aureus*: This is a pimple-like infection of the skin and hair follicles. Milk affected goats last and use individual towels to disinfect udder.
- Furunculosis: This is a severe form of *Staph aureus*. Boil-like lesions on the teats will rupture and drain. The doe may initially develop a fever and go off feed before the lesions develop. Isolate from the rest of the does until the lesions stop draining. Penicillin may be helpful at early stages.

Urolithiasis in Bucks or Castrates

Cause: Urinary calculi occur when crystals composed of organic matter and minerals form stones within the urinary tract. These stones cause trauma to the bladder and to the urethra. The stones can become lodged in the urethra or the urethra process, blocking urine flow. The formation of these stones can be related to the diet, or some family lines are predisposed to certain forms of stones.

Clinical signs: The most common clinical sign is an inability to pass urine. The buck will strain and make exaggerated motions trying to pass urine and show pain. The hairs on the skin surrounding the penis will be dry and may have small stones attached to them. If the bladder ruptures, the animal will appear to be more comfortable but later go off feed, and his abdomen will swell with fluid.

If you notice these signs it is advisable to call your veterinarian. You can set the buck on its rump and examine the penis by pushing it out of the prepuce. When the buck is on his rump it is a natural tendency for the penis to protrude. There is a small wormlike structure at the end of the penis called the urethral process. This is a site where the obstruction is most likely to occur. This process will be swollen and enlarged, and on palpation small stones can be felt.

Treatment: Your veterinarian should be called. Your veterinarian will discuss treatment options available to you.

Prevention: The goal is to dissolve any stones present and to prevent any more from forming. The buck (or the group if it is a group problem) should have free access to loose salt and fresh palatable water. Salt can be directly added to the feed so that the goat is receiving 60 to 100 gm per day. The salt should be added gradually over a period of several days.

Ammonium chloride can be added to the feed at a level of .5 to 1 percent of the goats daily dry matter intake. Ammonium chloride should only be fed to the group or individual that are affected. Ammonium chloride can be added by veterinary prescription into a prepared feed at a feed mill.

The diet should also be investigated to see if there is any predisposing cause, such as high calcium, oxalate, or silicate in the diet. Often the dietary treatment will be dependent on the type of feed available. Because there may be a genetic component, the buck's family history should be explored.

Common Parasites And Their Control

Internal and external parasites can be a problem in any livestock unit. Fortunately, in goats parasites are relatively easy to control because of the goats fastidious nature and their grazing behavior. See section on *Drugs approved for sheep and goats* for further information on treatment of parasites.

External Parasites (parasites that live on the surface of the body)

Lice

There are both blood sucking and biting (live off scurf) lice. Transmission is by contact. They are seen more often in the winter and are found around the tail, head, and the neck. They tend to blend in with the skin and may be difficult to see. Large numbers reflect the health status of the goat and are associated with weight loss. They can drag the animal down if the infestation is heavy. Debility and anemia are associated with sucking lice. Itching and hair loss are the main clinical signs of biting lice. For control use rotenone dust or cat flea spray every seven to ten days. Most products cannot be used in lactating does, so check product labels.

Sarcoptic mange

This mite breaks through the skin and dines on tissue fluids. Lesions appear first around the eyes and ears and spread from there. Itchy thickening of the skin and moist eczema with crusting are the main clinical features. Deep skin scrapings around the margin of the lesions can be examined under a microscope for the mites. Ivomec seems to be the treatment of choice, but it cannot be used in lactating does.

Chorioptic mange

These mites live and eat debris on the skin surface. Clinical signs depend on the number and the environmental temperature, with the highest numbers seen during cold weather. Crusty lesions are confined to the lower parts of the legs, primarily behind the fetlocks. No real symptoms are associated with the mite except itchiness. This parasite can live outside the host for periods up to ten weeks.

Demodectic mange

Although present in the hair follicles, these mites do not seem to be a problem. Stress seems to favor development of lesions. Multiple small nodules appear in the skin. The mites can be expressed from the skin nodule.

Psoroptic mange

This occurs on the inner surface of the ear and is associated with head shaking. The prevalence is high; and they have been found in the ears of kids as young as ten days old. Two injections of Ivomec (0.2mg/kgBW) two weeks apart provides control.

Internal Parasites (parasites that live within the body, usually the intestinal tract)

Coccidia

These are small single-celled parasites (protozoa) that live most of their lives within the host, but part of their life is spent outside of the host in the manure. These parasites are commonly called coccidia and although many families exist, only a few families cause trouble:
Toxoplasma are found in the feces of cats, and when eaten by goats can cause abortions; these parasites also infect human brains.
Cryptosporidia cause scours in kids less than two weeks of age and mild to severe

gastrointestinal symptoms in humans.
Eimeria species cause scours or bloody scours in kids over three weeks of age. This disease we all know as coccidiosis.

Manure builds up in the environment in areas where mature goats congregate or live. The nursing kids must eat the manure from the adult goats in order for their rumen and intestinal tracts to adapt to forage and grain diet. From our perspective this is a disgusting habit, but from a kid's perspective it is essential to life.

Life Cycle of Coccidia

Let's follow this single cell parasite as it journeys through its rather short life.

"Hi, I am a single-cell parasite called an *Eimeria oocyst*, and my life story would make the ultimate in microscopic horror movies. I, along with many of my family, have just escaped from the intestines and are now living free within a fecal pellet. To you this may not seem to be a very glamorous life, but to me life is just beginning. How many of you can say that you are resistant to environmental factors and can survive for up to a year sheltered only by feces? I can even tell when spring and warmer weather approaches because strange thing begin to happen to my body. Over a period of three to five days I start to divide and begin to lose my identity and become eight sporzoites; my life is further disrupted when a kid eats me. Once in the intestinal tract, my new body looks for a safe place to develop further and thus enters an intestinal epithelial cell, where, as a *sporzoite*, my body undergoes another change into a *schizont*. I then develop and eventually rupture, releasing thousands of little *merozoites*, which finally destroy the epithelial cell they called home.

The last generation of merozoites enters into another epithelial cell. My sex life begins when one develops into a male and the other a female gametocyte. When the female is fertilized, I am released from the cell, and my quiet life of an *oocyte* begins again. Once the cell is destroyed, it leaves a small wound, which exposes the sensitive tissues under the cell. If enough cells are destroyed, serum and fluids and often blood can drain from the wound, and the raw area can no longer carry out its normal function, so the kid starts to scour. The amounts of damage that I can do depends on how many of my sibling oocytes are present in the feces that a kid ingests.

If our numbers are few, we will only stimulate the immune system. If our numbers are large, we can cause clinical scours, dehydration, anemia, and, if severe enough, death. In most healthy kids the scouring will be mild and last for four to seven days; some kids will require oral or intravenous electrolytes to correct dehydration, and a few kids may require a blood transfusion, but such kids are already compromised by the lack of colostrum. Fortunately most kids survive; the intestinal epithelium heals in four to seven days, and the immune system is stimulated to reduce the impact of future invasions. But can I be stopped? I can, but I am not suicidal, so if you want to find out where I am vulnerable, you will have to ask your veterinarian."

Prevention

The key to preventing coccidiosis is to reduce environmental contamination. You can do this by keeping the pens clean and well bedded and to be sure your does kid and nurse in a clean area, free from the winter build up of manure. A cocccidiostat, which kills the merozoites life form of the coccidia, can be fed to the does three to four weeks prior to kidding to reduce the number of oocytes shed into the clean environment.

Table 21 A Summary of the Coccidiostats Available

Coccidiostat	Dose	Comments
Baycox	Oral 10 mg/kg BW on two subsequent days	Decreases oocytes in feces for 3 weeks; by 5 weeks large numbers of oocytes in feces
Monensin	Mixed in feed at 11 to 15 mg per ton, usually in a creep ration	Reduces shedding by half within 6 days. Monensin is toxic to dogs and horses and is not palatable if improperly mixed.
Decox	22 mg per kg BW in feed	Can feed in the creep ration and grower ration
Bovetec	11 to 33 mg/ton in feed	Not as effective as others
Amprolium	25 to 5O mg/kg BW in feed or water	Feed for 2 weeks to several months of age. Amprolium can have a negative effect on thiamine and predispose the goats to polioencephalomalasia

Nematodes, Round Worms, or Helminths:
Life Cycle from the Parasites' Perspective

"What a life—sheltered within the intestinal tract and surrounded by nutrients that I share with, or in some cases steal from, my host or hostesses. A number of parasite families have selected their favorite site to live; some select the abomasums; others live at various sites along the intestinal tract; some actually bury their heads into the intestinal wall and suck blood from the host. Most of the helminths are egg-laying machines, except for one family that lays a few very large eggs. I never really get to see my children because the eggs I lay are passed in the feces to the outside world. I don't remember much about my life outside the body, but I did interview a young adult worm, which had just come from the outside. The following are some highlights of her adventure."

"I really don't remember my life as an egg; once I was passed in the feces, I quickly turned into a first-stage larva. Lots of tasty bacteria from the feces were present to nurture my rapidly growing body. Soon I started to scratch, and I molted and appeared as a second-stage larva. And now my food is running out and I can feel another molt coming on.

"Wow that was a rough one. Now I have a sudden urge to return to the warm smorgasbord, so I will climb up this blade of grass and wait for a grazing doe to come along and eat me. I will then leisurely make my way through her lungs and liver until I reach her intestinal tract. If the time of year is right and she is in mid-pregnancy, I will rest a while until her noisy kids wake me up. Then I will resume my journey to her intestinal tract. In three to four weeks my juvenile stage will be over, and I, along with my siblings, will start to lay thousands of eggs, contributing to maintenance of our population.

"This sounds like an idyllic life for a worm, but nothing is farther from the truth. Thousands of my siblings can perish during the first two larval stages if trapped in a fecal pellet desiccated by the sun, unless moisture comes to soften the pellet and release them. Those that manage to free themselves and become infective develop a hard cuticle. If adverse conditions occur they can live on stored energy reserves, depending on the ambient temperature. If conditions become worse, they can burrow into the soil, which further protects them. Upon hatching, all the larvae wish for hot moist weather."

Treatment (Deworming Strategies)

Even migrating through the warm, nurturing body can be perilous for worms, especially if the flock is on a strategic deworming program, when the goats are dewormed while the larva and worms are in the ewe and not in the environment.

Deworming strategies include:

- Pasture rotation[23]
- Deworming prior to going on pasture
- Keeping confinement facilities dry and well bedded
- Routine fecal checks for worm eggs
- Deworming 3 to 4 weeks before kidding
- Cutting the feed intake by half the day before deworming to allow greater exposure to the dewormer
- Selecting the dose of dewormer based on the heaviest animal in the group

Eventually the doe's immune system takes over and keeps things in check.

Gone are the days of "hit and miss" deworming, putting goats out on the same over-grazed native or poorly managed tame pasture year after year, and under-dosing with dewormer.

Chapter 6

The Responsible Use of Medications

Introduction

Caution: Always check with your veterinarian before using pharmaceuticals because the product name, dosage, and withdrawal times may be different than presented here.

A limited number of drugs and vaccines are approved for goats. Part of this is related to the high cost for drug companies to study the effect of a new drug according to the guidelines of Health Canada's Veterinary Drug Directorate (VDD).The Canadian Food Inspection Agency (CFIA) Veterinary Biologics is responsible for vaccine approval. Often the veterinarian and the producer assume that if the product is approved for cattle or sheep, it is safe for goats. Pharmaceuticals that have on the label or in the instructions "for goats/livestock" are the only ones tested for use in goats. Since changes in a drugs status occur frequently, I would advise discussing current drug therapies with your veterinarian.

The following definitions should help the producer understand their responsibilities and those of their veterinarian.

Drugs approved for Sheep and Goats

Prescription Drugs: "Schedule F Drugs"

This is a product that can only be sold to a livestock producer by a veterinarian. To use a "schedule F drug," you must be the veterinarian's client, and the vet must have examined the animal prior to prescribing the drug. In addition, your veterinarian should only dispense enough medication to treat that animal or group of animals. Your veterinarian is responsible for following these guidelines. A prescription drug can be identified by the symbol "Rx" after the product name.

Prescription Feeds

These are medicated feeds manufactured according to a written prescription supplied by a veterinarian. These feeds are used by a client of the veterinarian for a specific purpose for the control of disease or enhancement of production. Such feeds are regulated by the Animal Health and Protection Agency of the CFIA.

Over the Counter Drugs (OTC)

These products can be purchased from any store that specializes in livestock supplies or from your veterinarian. You are responsible for reading the label recommendations and following the exact dosage, instructions for administration, and administering it to the approved animals as indicated on the label.

For Veterinary Use Only

These are prescription or OTC drugs that can only be administered to animals.

"Extra Label" Drug Use (ELDU)

This is a very important concept for the producer to understand. This term applies to circumstances in which there is no approved animal drug in the required dosage, formulation, or concentration. A veterinarian is allowed to use extra label drugs within the context of a valid vet/client/patient relationship. If the product is purchased OTC, the producer assumes the responsibility for following the label instructions exactly. If the drug in question is a prescription, the veterinarian is responsible for explaining the label instructions and how any deviation from the label instructions should be handled. This is particularly important in establishing the withdrawal period for slaughter and milk. If the producer recognizes a deviation from the label, the veterinarian should be questioned. In food-producing animals, extra label use is limited to cases in which the animal's health is threatened or suffering or when death may result from the failure to treat with the drug in question. In the United States the regulations related to "extra label" use are being tightened. If those changes are approved, further restrictions will be placed on the extra label use of drugs.

Withdrawal Period

This is the length of time that it takes for the last administered dose of a drug to be eliminated from the body. The withdrawal period is only valid if the exact label dosage and route of administration are followed and the approved species is treated. If any of these instructions are violated, the withdrawal time on the label is no longer valid.

Case Studies: Consequences of Off Label Administration of OTC drugs

Case 1

A dairy farmer is told by her veterinarian that penicillin, an OTC drug, can be injected under the skin rather than in the muscle. The producer follows the label dosage but injects it under the skin, and she withholds the milk for the required time period. To the farmer's surprise, she is penalized for shipping milk that is positive for penicillin.

Case 2

Another example of a serious nature occurred after a sheep producer used 20 cc of penicillin in the hind leg muscle of his feedlot lambs as a treatment for foot rot. He did not slaughter these lambs until the end of the label withdrawal period. One Sunday the family was enjoying a roasted leg of lamb when their son, who was allergic to penicillin, collapsed. Since this had happened before, the producer and his wife were quick to respond and saved their son's life.

Expiration Date

This date is on the label of most antibiotics, vaccines, vitamins, and other products that deteriorate with age. The drug's expiration date is determined under laboratory conditions, not under variable farm conditions. Some products may even break down into harmful components when this date is passed.

Caution: Always check the expiration date before you purchase a product. If the product is on sale because the date is approaching, be prepared to use it immediately.

Some deals may not be as good as they seem. Some of the products may be effective and safe after the date indicated, but on the other hand, using it may be a costly mistake on your part.

Often drugs kept on the farm are not maintained under ideal conditions (for example, needles are left in the tops; products are exposed to extreme environmental conditions, or repeated use has introduced bacteria which, if injected, may be harmful).

The Responsible Use of Medications

Always check your drugs on a regular basis and discard any that have expired, were accidentally frozen or exposed to excess heat, or if the tops have been ruined by repeated use. With any pharmaceuticals, the bottom line is to protect our environment and our food supply from drug residues.[24] In the case of antibiotic and anthelmentics, you want to prevent the possibility of developing drug resistant bacteria and parasites. Therefore:

- Follow directions closely. If you have any concerns, phone your veterinarian.
- Read the label closely, noting dosage and withholding times.
- Store the drug according to label instructions.
- If treating or vaccinating a group of animals, purchase only enough for that particular use.
- If you are unsure of the withdrawal time, extend this period. I recommend a minimum of thirty days if unknown.
- Always identify the animal to be treated so you do not treat the wrong one.
- Keep individual records of individual medication, dose, and date treated.

Antimicrobials for Sheep and Goats

Antimicrobials are used to treat bacterial and viral and fungal diseases, to control and prevent infection, for growth promotion, and to improve feed efficiency. Individual animals can be treated, or the medication can be put in the water or feed for groups of animals at a therapeutic or prophylactic (preventive) concentration.

Vaccines (Biologicals)

Vaccines are used to protect a goat against bacterial and viral diseases common to your area. Generally the recommendation is to vaccinate kids at two to three months of age and then six weeks later. If the goat is maintained in the herd, an annual vaccination may be recommended. The need for the annual vaccination is debatable. If the organism is common in the environment and the goat is constantly exposed, once protected by the first annual vaccination, natural exposure will stimulate the immune system, making an annual vaccine unnecessary.

Controlling External Parasites in Goats

When you look for an appropriate product, check the label for the words "goats or livestock." If you do not find these words, read the caution statement to make sure goats do not react adversely to the product. All such products, if mishandled, can be harmful to you and your goats. If the label states that "before use, every inch of your body must be protected," that's exactly what you must do.
Caution: Do not assume that a product approved for sheep is safe to administer to goats. For small ruminants, always read the label and caution statements.

These products and their names change frequently, so the following list might be outdated as I type.

Table 22: Products Used in Goats to Control Lice and Other External Parasites

Product Name	Active Ingredient	Comments
Co-op Louse powder	Rotenone	For sheep and goats
Dri Kil"	Rotenone Sulphur	
Dusting powder	Carbaryl	
Echiban 25 Fly Killer	Permethrim	Horn flies, face flies, black flies, and mosquitoes 90 day withdrawal for meat
Products Not Approved for Goats		
Fly Ear Tags	Variety	Not approved: Have been cut in half and used to control face flies and bots

Note: *Pour on products* such as Spot-on and other organophosphates are not approved. If you use them off label, the dose must be carefully measured. They should not be used in stressed or diseased goats or kids under three months of age. Withdrawal period for cattle is forty-five days.

Table 23 Vaccines Approved for Use in Sheep**

Product Name	Dose (ml or cc)	Route of administration*	Withdrawal days**	Purpose /Best Time to Use
Anthrax spore vaccine	1 ml	SQ	42	If anthrax is a problem, vaccinate annually 4 weeks prior to potential disease outbreak. Revaccinate in outbreak after 1st vaccine in 2-3 weeks
Campylobacter fetus bacterin-0vine)	5 ml	SQ behind the elbow	21	1st vaccine before breeding Repeat in 60 to 90 days and then annually
Case –Bac	2	SQ, repeat in 2-4 weeks then annually	21	Caseous lymphadenitis
Caseous D-T	2	SQ, repeat 4 weeks later and then annually	21	For tetanus and CL and clostridium perfringins Type D
Chlamydia Psittaci bacterin	2	SQ Upper neck 60d prior to breeding, repeat in 30 days and then annually prior to breeding	60	Ovine enzootic abortion
Covexin 8	4, followed by 2 ml 6 wks later	SQ	21	8 way for Clostridial diseases
Defensor 3	2	IM, repeat annually	21	For rabies
Imrab 3	2	SQ or IM	21	For rabies
Imrab Large Animal	2.5	SQ or IM	21	For rabies
Prorab	2	SQ	21	For rabies
Tsvax 7	4, and then 2 ml 6 weeks later	SQ	42	Clostridial diseases
Tasvax 8	4, and then 2ml 6 weeks later	SQ	21	Clostridial diseases
Ultrachoice 7	1	SQ	21	Clostridial diseases
Ultrachoice 8	1	SQ	21	Clostridial diseases
Vision 7	2	SQ	21	Clostridial diseases

Vision 8	2	SQ,	21	Clostridial diseases
Vision CD-T	1	IM	21	Vaccine for tetanus

**** Can be used in goats only under veterinary supervision and advise about withdrawal times. Adverse reaction may occur as the vaccines have not been tested in *goats***

"If ***you deworm your goats and get a response, you have waited too long***."

Table 24 Dewormers Approved For Use in Canadian Sheep/Goats

			Effective Against				
Name	Route	WD	Nose Bots	Intestinal Parasites	Tape Worm	Lung Worm	Liver Flukes
Ivomec Inject	SQ has some use against Keds and sucking lice	35 d Not for milk sheep	+	+	-	+	-
Ivomec* Drench	Oral	14 d	+	+	-	+	-

Table 25 Products Available for Cattle but Not Approved for Sheep/Goats (must be prescribed by a veterinarian)							
Valbazen** Albendazole†	Oral 10 mg/kg BW	> 30 d	?	+	+	+	+ 4.8 mg/kg BW
Panacur Fenbendazole†	Oral 5 mg/kg BW	14 d, use > 30d	?	+	+	+	+
Telmin Drench Membendazole + Trichlor-fon**	Oral 10 mg/kg BW	Equine Prep. > 30 d	+?	+	+	+	-
Safe Guard Fenbendazole†	Oral	5 mg/kg BW	?	Yes	10 mg/ kg BW	Yes	
Exhelm E Morental Tar-trate†	Oral 10 mg/kg BW	30 d Cattle	?	+	-	+	-

* Reduce feed intake by half twenty-four hours prior to drenching. The reduction in feed slows passage through gastrointestinal tract

For more information on the Canadian Sheep and Lamb Food-Safe Farm Practices, call 1-888-684-7739, e-mail france@cansheep.ca, or visit the new FSFP Web site for online training at http://fsfp.cansheep.ca

Chapter 7

Conclusion

Finally this book is complete. I recognize that I may not have answered all of your questions, but hopefully I have answered some. Unfortunately, because of the Internet, cell phones, and a glut of information, some of what I have written will already be outdated or not be supported by the information in cyberspace.

Some of the information not covered can be found within the references, which include links to Web sites. Although I have not covered organic farming, it's not because I don't support organic farming. As a matter of fact, I support any agricultural operation that is sustainable through proper land use, humane livestock production, and the responsible use of antibiotics and other agricultural chemicals. Organic farming is the ideal, but with the current proposed regulations, is it truly practical under today's conditions?[25]

Goats are truly amazing animals that can be raised under many different management systems and for many purposes. They are intelligent, and if properly cared for will reap you and your family many benefits. I wish you all the very best in your ventures. I hope this book will set you on the right track.

Endnotes

1. Cud chewing is essential for ruminant digestion. The rumen holds the feed as they eat. During a quiet period, a bolus of rumen content is regurgitated, and the fiber in the feed is further broken down by the teeth, making more surface area for the rumen micro flora to invade and digest.
2. Hay and sampling equipment can be leased or obtained from a feed-testing laboratory, local agricultural office, large animal veterinarian, or local feed company. If you are buying feeds from a local mill, they often will come and sample your feeds for you.
3. Jim Sprinkle & Rob Grumbles, Nutritional characteristics of Arizona Browse. http://ag.arizona.edu/pubs/animal/az1273/ Retrieved January 26, 2010.
4. 2006, ASI, A. Peischel and D.D. Henry, Jr., Targeted Grazing: A Natural Approach to Vegetation Management and Landscape Enhancement. www.cnr.uidaho.edu/rx-grazing/Handbook.htm/ Retrieved January 26, 2010.
5. TDN (Total Digestible Nutrients) is a measure of the amount of energy in the feed. I find it the easiest measure to use (1 kg TDN = 4.409 megacalories of digestible energy).
6. DE, like TDN, is a measure of energy. DE represents the amount of energy taken in by the body when the food is digested. I use DE energy when milk is the main diet.
7. Canadian National Goat Federation, Traceability and National ID www.cangoats.com/en/traceability.html/ Retrieved January 26, 2010. Alberta Goat Breeders Association, News Archive www.albertagoatbreeders.ca/default.asp?contentID=710/ Retrieved January 26, 2010.
8. Rick Machen, Uvalde Eddie Holland, & Warren Thigpen, Disbudding Kid Goats—Is It Commercially Applicable? www.goatworld.com/articles/disbudding/disbudding.shtml/ Retrieved January 26, 2010.
9. Robin L. Walters, How To Tube Feed A Kid Goat www.goatworld.com/articles/kidding/tubefeedingrw.shtml/ Retrieved January 26, 2010.
10. Hypothermic kids have a very low body temperature. This usually occurs in the first 3 to 4 days after birth and is associated with failure to nurse and starvation.
11. Robert Spencer, Meat Goat Quality Assurance Through Proper Injection Sites http://motesclearcreekfarms.com/asp/articles/meat-goat-quality-assurance.asp Retrieved January 26, 2010.
12. Vicnet, a division of the State Library of Victoria, Australia, Hoof Trimming http://home.vicnet.net.au/~goats/dgsavictoria/hoof_trimming.htm Retrieved January 26, 2010.
13. Diagnosis of Internal Parasitism in Goats www2.luresext.edu/goats/library/fec.html Retrieved January 26, 2010..
14. E (Kika) de la Garza Institute for Goat Research www2.luresext.edu/goats/index.htm Retrieved January 26, 2010.
15. Sheep and Goat Fecal Analysis www.microscope-microscope.org/applications/animals/fecal_analysis.htm Retrieved January 26, 2010.
16. P. Chemineau, G. Baril, B. Leboeuf, M.C.Maurel & Y. Cognie, Recent Advances in the Control of Goat Reproduction http://ressources.ciheam.org/om/pdf/c25/97605952.pdf/ Retrieved January 26, 2010.
17. Jean-Marie Luginbuhl, Preparing Meat Goats For The Breeding Season, ANS OO-6O2MG
18. Sheep and Goat.com—General Goat Links www.sheepandgoat.com/goatlnks.html Retrieved January 26, 2010.

19. Chemineau, G. Baril, B. Leboeuf, M.C. Maurel, F. Roy, M.Pellicer-Rubio, B. Malpaux, Y. Cognie. Recent advances in goat reproduction, P. Anim., 12(2), 135–146. INRA
20. Meiosis—Reproduction in the Goat www.imagecyte.com/chromo.html/ Retrieved January 26, 2010.
21. Scrapie http://ext-search-recherche.inspection.gc.ca/search?q=scrapie&site=ex_global&client=active_e&proxystylesheet=active_e&output=xml_no_dtd&hl=fr&ie=latin1&oe=latin1&access=p&ip=192.168.121.1&btnG= percentA0Search percentA0
22. Scrapie www.aphis.usda.gov/animal_health/animal_diseases/scrapie/
23. Peter Stockdale. Living with Worms in Organic Sheep Production. 2008 Canadian Organic Growers Inc. *Practical Skills Handbook.*
24. Health Canada, Release of the Final Report of the Advisory Committee on Animal Uses of Antimicrobials and Impact on Resistance and Human Health. www.hc-sc.gc.ca/dhp-mps/pubs/vet/amr-ram_final_report-rapport_06-27_tc-tm-eng.php/ Retrieved January 26, 2010.
25. Forge, Frédéric (2001, February 19). Organic Farming in Canada, Government of Canada, Parliamentary Research Branch, Shttp://www.aphis.usda.gov/animal_health/animal_diseases/scrapie/science and Technology Division http://dsp-psd.tpsgc.gc.ca/Collection-R/LoPBdP/BP/prb0029-e.htm/ Retrieved January 26, 2010. In the USA: Organic Agriculture www.ers.usda.gov/Briefing/Organic/ Retrieved January 26, 2010. World-Wide Opportunities on Organic Farms, USA www.wwoofusa.org/ Retrieved January 26, 2010.

he United States
asters